Studies of Trees

Jacob Joshua Levison

Alpha Editions

This edition published in 2024

ISBN : 9789364739306

Design and Setting By
Alpha Editions
www.alphaedis.com
Email - info@alphaedis.com

As per information held with us this book is in Public Domain.
This book is a reproduction of an important historical work. Alpha Editions uses the best technology to reproduce historical work in the same manner it was first published to preserve its original nature. Any marks or number seen are left intentionally to preserve its true form.

CONTENTS

Preface .... - 1 -

Introduction .... - 3 -

Chapter I .... - 4 -

    How To Identify Trees .... - 4 -

Chapter II .... - 21 -

    How To Identify Trees—(*Continued*) .... - 21 -

Chapter III .... - 72 -

    How To Identify Trees—(*Continued*) .... - 72 -

Chapter IV .... - 92 -

    The Structure and Requirements of Trees .... - 92 -

Chapter V .... - 101 -

    What Trees to Plant and How .... - 101 -

Chapter VI .... - 111 -

    The Care of Trees .... - 111 -

Chapter VII .... - 144 -

    Forestry .... - 144 -

Chapter VIII ....- 172 -

　Our Common Woods: Their
　Identification, Properties and Uses ....- 172 -

Chapter IX ....- 188 -

　An Outdoor Lesson on Trees ....- 188 -

PREFACE

In presenting this volume, the author is aware that there are several excellent books, dealing with one phase or another of tree life, already before the public. It is believed, however, that there is still need for an all-round book, adapted to the beginner, which gives in a brief and not too technical way the most important facts concerning the identification, structure and uses of our more common trees, and which considers their habits, enemies and care both when growing alone and when growing in groups or forests.

In the chapters on the identification of trees, the aim has been to bring before the student only such characters and facts as shall help him to distinguish the tree readily during all seasons of the year. Special stress is laid in each case on the most striking peculiarities. Possible confusion with other trees of similar appearance is prevented as far as possible through comparisons with trees of like form or habit.

Only such information is given concerning the structure and requirements of trees as will enable the reader better to understand the subsequent chapters. In the second half of the book, practical application is made of the student's general knowledge thus acquired, and he is acquainted with the fundamental principles of planting, care, forestry, wood identification and nature study.

The author recognizes the vastness of the field he is attempting to cover and the impossibility of even touching, in a small hand-book of this character, on every phase of tree study. He presumes no further; yet he hopes that by adhering to what is salient and by eliminating the less important, though possibly interesting, facts, he is able to offer a general and elementary *résumé* of the whole subject of value to students, private owners, farmers and teachers.

In the preparation of Chapter VIII on "Our Common Woods: Their Identification, Properties and Uses," considerable aid has been received from Prof. Samuel J. Record, author of "Economic Woods of the United States." Acknowledgment is also due to the U. S. Forest Service for the photographs used in Figs. 18, 122 to 138 inclusive and 142; to Dr. George B. Sudworth, Dendrologist of the U. S. Forest Service, for checking up the nomenclature in the lists of trees under Chapter V; to Dr. E. P. Felt, Entomologist of the State of New York, for suggestions in the preparation of the section of the book relating to insects;

to Dr. W. A. Murrill, Assistant Director of the New York Botanical Gardens, for Fig. 108; and to Mr. Hermann W. Merkel, Chief Forester of the New York Zoological Park, for Figs. 26, 59 and 60.

J. J. LEVISON.

BROOKLYN, N. Y.

June, 1914.

Introduction

A good many popular books on trees have been published in the United States in recent years. The continually increasing demand for books of this character indicates the growing public interest not only in the trees that we pass in our daily walks, but also in the forest considered as a community of trees, because of its æsthetic and protective value and its usefulness as a source of important economic products.

As a nation, we are thinking more about trees and woods than we were wont to do in the years gone by. We are growing to love the trees and forests as we turn more and more to outdoor life for recreation and sport. In our ramblings along shady streets, through grassy parks, over wooded valleys, and in mountain wildernesses we find that much more than formerly we are asking ourselves what are these trees, what are the leaf, flower, twig, wood and habit characteristics which distinguish them from other trees; how large do they grow; under what conditions of soil and climate do they thrive best; what are their enemies and how can they be overcome; what is their value for wood and other useful products; what is their protective value; are they useful for planting along streets and in parks and in regenerating forests; how can the trees of our streets and lawns be preserved and repaired as they begin to fail from old age or other causes? All these questions and many more relating to the important native and exotic trees commonly found in the states east of the Great Lakes and north of Maryland Mr. Levison has briefly answered in this book. The author's training as a forester and his experience as a professional arboriculturist has peculiarly fitted him to speak in an authoritative and interesting way about trees and woods.

The value of this book is not in new knowledge, but in the simple statement of the most important facts relating to some of our common trees, individually and collectively considered. A knowledge of trees and forests adds vastly to the pleasures of outdoor life. The more we study trees and the more intimate our knowledge of the forest as a unit of vegetation in which each tree, each flower, each animal and insect has its part to play in the complete structure, the greater will be our admiration of the wonderful beauty and variety exhibited in the trees and woods about us.

J. W. Toumey,
Director, Yale University Forest School.

New Haven, Conn.,

June, 1914.

Chapter I

How To Identify Trees

There are many ways in which the problem of identifying trees may be approached. The majority attempt to recognize trees by their leaf characters. Leaf characters, however, do not differentiate the trees during the other half of the year when they are bare. In this chapter the characterizations are based, as far as possible, on peculiarities that are evident all year round. In almost every tree there is some one trait that marks its individuality and separates it, at a glance, from all other trees. It may be the general form of the tree, its mode of branching, bark, bud or fruit. It may be some variation in color, or, in case of the evergreen trees, it may be the number and position of the needles or leaves. The species included in the following pages have thus been arranged in groups based on these permanent characters. The individual species are further described by a distinguishing paragraph in which the main character of the tree is emphasized in heavy type.

The last paragraph under each species is also important because it classifies all related species and distinguishes those that are liable to be confused with the particular tree under consideration.

GROUP I. THE PINES

FIG. 1.—Twig of the Austrian Pine.

How to tell them from other trees: The pines belong to the *coniferous* class of trees; that is, trees which bear cones. The pines may be told from the other coniferous trees by their leaves, which are in the form of *needles* two inches or more in length. These needles keep green throughout the entire year. This is characteristic of all coniferous trees, except the larch and cypress, which shed their leaves in winter.

FIG. 2.—Twig of the White Pine.

The pines are widely distributed throughout the Northern Hemisphere, and include about 80 distinct species with over 600 varieties. The species enumerated here are especially common in the eastern part of the United states, growing either native in the forest or under cultivation in the parks. The pines form a very important class of timber trees, and produce beautiful effects when planted in groups in the parks.

How to tell them from each other: The pine needles are arranged in *clusters*; see Fig. 1. Each species has a certain characteristic number of needles to the cluster and this fact generally provides the simplest and most direct way of distinguishing the different pines.

In the white pine there are *five* needles to each cluster, in the pitch pine *three*, and in the Scotch pine *two*. The Austrian pine also has two needles to the cluster, but the difference in size and character of the needles will distinguish this species from the Scotch pine.

THE WHITE PINE (*Pinus strobus*)

Distinguishing characters: The tree can be told at close range by the number of needles to each cluster, Fig. 2. There are **five** needles to each cluster of the white pine. They are bluish green, slender, and about four inches in length.

At a distance the tree may be told by the **right angles** which the branches form with the main trunk, Fig. 3. No other pine shows this character.

Form and size: A tall tree, the stateliest of the evergreens.

Range: Eastern North America.

Soil and location: Prefers a deep, sandy soil, but will grow in almost any soil.

Enemies: Sucking insects forming white downy patches on the bark and twigs, the *white pine weevil*, a boring insect, and the *white pine blister rust*, a fungus, are among its principal enemies.

FIG. 3.—The White Pine.

Value for planting: Aside from its value as an ornamental tree, the white pine is an excellent tree to plant on abandoned farms and for woodlands and windbreaks throughout the New England States, New York, Pennsylvania, and the Lake States.

Commercial value: The wood is easily worked, light, durable, and will not warp. It is used for naval construction, lumber, shingles, laths, interior finish, wooden ware, etc.

Other characters: The *fruit* is a cone, four to six inches long.

Comparisons: The tree is apt to be confused with the *Bhotan pine* (*Pinus excelsa*), which is commonly grown as an ornamental tree. The Bhotan pine, however, has needles much longer and more drooping in appearance.

THE PITCH PINE (*Pinus rigida*)

Distinguishing characters: Here there are **three** needles to each cluster, Fig. 4. They are dark, yellowish-green needles about four inches long. The rough-looking *branches* of the tree may be seen *studded with cones* throughout the year, and *clusters of leaves* may be seen *sprouting directly from the trunk* of the tree; see Fig. 5. The last two are very characteristic and will distinguish the tree at a glance.

Form and size: It is a low tree of uncertain habit and extremely rough looking at every stage of its life. It is constantly full of dead branches and old cones which persist on the tree throughout the year.

Range: Eastern United States.

Soil and location: Grows in the poorest and sandiest soils where few other trees will grow. In New Jersey and on Long Island where it is native, it proves so hardy and persistent that it often forms pure stands excluding other trees.

FIG. 4.—Twig of the Pitch Pine.

Enemies: None of importance.

Value for planting: Well adapted for the sea coast and other exposed places. It is of extremely uncertain habit and is subject to the loss of the lower limbs. It frequently presents a certain picturesqueness of outline, but it could not be used as a specimen tree on the lawn.

FIG. 5.—The Pitch Pine.

Commercial value: The wood is coarse grained and is used for rough lumber, fuel, and charcoal.

Other characters: The *fruit* is a cone one to three inches long, persistent on the tree for several years.

THE SCOTCH PINE (*Pinus sylvestris*)

Distinguishing characters: There are **two** needles to each cluster, and these are *short* compared with those of the white pine, and *slightly twisted*; see Fig. 6. The *bark*, especially along the upper portion of the trunk, *is reddish* in color.

Form and size: A medium-sized tree with a short crown.

Range: Europe, Asia, and eastern United States.

Soil and location: Will do best on a deep, rich, sandy soil, but will also grow on a dry, porous soil.

Enemies: In Europe the Scotch pine has several insect enemies, but in America it appears to be free from injury.

Value for planting: Suitable for windbreaks and woodland planting. Many excellent specimens may also be found in our parks.

Commercial value: In the United States, the wood is chiefly used for fuel, though slightly used for barrels, boxes, and carpentry. In Europe, the Scotch pine is an important timber tree.

Comparisons: The Scotch pine is apt to be confused with the *Austrian pine* (*Pinus austriaca*), because they both have two needles to each cluster. The needles of the Austrian pine, however, are much longer, coarser, straighter, and darker than those of the Scotch pine; Fig. 1. The form of the Austrian pine, too, is more symmetrical and compact.

FIG. 6.—Twig of the Scotch Pine.

The *red pine* (*Pinus resinosa*) is another tree that has two needles to each cluster, but these are much longer than those of the Scotch pine (five to six inches) and are straighter. The bark, which is reddish in color, also differentiates the red pine from the Austrian pine. The position of the cones on the red pine, which point outward and downward at maturity, will also help to distinguish this tree from the Scotch and the Austrian varieties.

GROUP II. THE SPRUCE AND HEMLOCK

How to tell them from other trees: The spruce and hemlock belong to the evergreen class and may be told from the other trees by their *leaves*. The characteristic leaves of the spruce are shown in Fig. 9; those of the hemlock in Fig. 10. These are much shorter than the needles of the pines but are longer than the leaves of the red cedar or arbor vitae. They are neither arranged in clusters like those of the larch, nor in feathery layers like those of the cypress. They adhere to the tree throughout the year, while the leaves of the larch and cypress shed in the fall.

The spruces are pyramidal-shaped trees, with tall and tapering trunks, thickly covered with branches, forming a compact crown. They are widely distributed throughout the cold and temperate regions of the northern hemisphere, where they often form thick forests over extended areas.

There are eighteen recognized species of spruce. The Norway spruce has been chosen as a type for this group because it is so commonly planted in the northeastern part of the United States.

The hemlock is represented by seven species, confined to temperate North America, Japan, and Central and Western China.

FIG. 7.—The Norway Spruce.

How to tell them from each other: The needles and branches of the spruce are *coarse*; those of the hemlock are *flat and graceful*. The individual leaves of the spruce, Fig. 9, are four-sided and green or blue on the under side, while those of the hemlock, Fig. 10, are flat and are *marked by two white lines* on the under side.

THE NORWAY SPRUCE (*Picea excelsa*)

Distinguishing characters: The characteristic appearance of the full-grown tree is due to the **drooping branchlets** carried on **main branches which bend upward** (Fig. 7).

Leaf: The leaves are dark green in color and are *arranged spirally*, thus making the twigs coarser to the touch than the twigs of the hemlock or fir. In cross-section, the individual leaflet is quadrilateral, while that of the pine is triangular.

Form and size: A large tree with a straight, undivided trunk and a well-shaped, conical crown (Fig. 7).

Range: Northern Europe, Asia, northern North America.

Soil and location: Grows in cool, moist situations.

Enemies: The foliage of the spruce is sometimes affected by *red spider*, but is apt to be more seriously injured by drought, wind, and late frosts.

Value for planting: Commonly planted as an ornamental tree and for hedges. It does well for this purpose in a cool northern climate, but in the vicinity of New York City and further south it does not do as well, losing its lower branches at an early age, and becoming generally scraggly in appearance.

FIG. 8.—A Group of Hemlock.

Commercial value: The wood is light and soft and is used for construction timber, paper pulp, and fuel.

Other characters: The *fruit* is a large slender cone, four to seven inches long.

Comparisons: The *white spruce* (*Picea canadensis*) may be told from the Norway spruce by the whitish color on the under side of its leaves and the unpleasant, pungent odor emitted from the needles when bruised. The cones of the white spruce, about two inches long, are shorter than these of the Norway spruce, but are longer than those of the black spruce.

It is essentially a northern tree growing in all sorts of locations along the streams and on rocky mountain slopes as far north as the Arctic Sea and Alaska. It often appears as an ornamental tree as far south as New York and Pennsylvania.

The *black spruce* (*Picea mariana*) may be told from the other spruces by its small cone, which is usually only about one inch in length. In New England it seldom grows to as large a size as the other spruce trees.

It covers large areas in various parts of northern North America and grows to its largest size in Manitoba. The black spruce has little value as an ornamental tree.

The *Colorado blue spruce* (*Picea parryana* or *Picea pungens*) which is commonly used as an ornamental tree on lawns and in parks, can be told from the other spruces by its pale-blue or sage-green color and its sharp-pointed, coarse-feeling twigs. Its small size and sharp-pointed conical form are also characteristic.

It grows to a large size in Colorado and the Middle West. In the Eastern States and in northern Europe where it is planted as an ornamental tree, it is usually much smaller.

FIG. 9.—Twig of the Norway Spruce.

HEMLOCK (*Tsuga canadensis*)

Distinguishing characters: Its leaves are arranged in **flat layers**, giving a flat, horizontal and graceful appearance to the whole branch (Fig. 8). The individual leaves are dark green above, lighter colored below, and are **marked by two white lines on the under side** (Fig. 10).

The leaves are arranged on little stalks, a characteristic that does not appear in the other evergreen trees.

Form and size: A large tree with a broad-based pyramidal head, and a trunk conspicuously tapering toward the apex. The branches extend almost to the ground.

Range: The hemlock is a northern tree, growing in Canada and the United States.

Soil and location: Grows on all sorts of soils, in the deepest woods as well as on high mountain slopes.

Enemies: None of importance.

Value for planting: The hemlock makes an excellent hedge because it retains its lowest branches and will stand shearing. In this respect it is preferable to the spruce. It makes a fair tree for the lawn and is especially desirable for underplanting in woodlands, where the shade from the surrounding trees is heavy. In this respect it is like the beech.

Commercial value: The wood is soft, brittle, and coarse-grained, and is therefore used mainly for coarse lumber. Its bark is so rich in tannin that it forms one of the chief commercial products of the tree.

Other characters: The *fruit* is a small cone about ¾ of an inch long, which generally hangs on the tree all winter.

Fig. 10.—Twig of the Hemlock.

GROUP III. THE RED CEDAR AND ARBOR-VITAE

How to tell them from other trees: The red cedar (juniper) and arbor-vitae may be told from other trees by their *leaves*, which remain on the tree and keep green throughout the entire year. These leaves differ from those of the other evergreens in being much shorter and of a distinctive shape as shown in Figs. 12 and 13. The trees themselves are much smaller than the other evergreens enumerated in this book. Altogether, there are thirty-five species of juniper recognized and four of arbor-vitae. The junipers are widely distributed over the northern hemisphere, from the Arctic region down to Mexico in the New World, and in northern Africa, China, and Japan in the Old World. The arbor-vitae is found in northeastern and northwestern America, China, and Japan. The species mentioned here are those commonly found in America.

How to tell them from each other: The *twigs* of the arbor-vitae are *flat and fan-like* as in Fig. 13; the twigs of the red cedar are *needle-shaped or scale-like* as in Fig. 12. The foliage of the arbor-vitae is of a lighter color than that of the red cedar, which is sombre green. The arbor-vitae will generally be found growing in moist locations, while the red cedar will grow in dry places as well. The arbor-vitae generally retains its lower branches in open places, while the branches of the red cedar start at some distance from the ground.

RED CEDAR (*Juniperus virginiana*)

FIG. 11.—The Red Cedar.

Distinguishing characters: The tree can best be told at a glance by its general form, size and leaves. It is a medium-sized tree with a *symmetrical, cone-like form*, Fig. 11, which, however, broadens out somewhat when the tree grows old. Its color throughout the year is dull green with a tinge of brownish red, and its bark peels in thin strips.

FIG. 12(*a*).—Twig of Young Cedar.

FIG. 12(*b*).—Twig of Cedar (Older Tree).

Leaf: In young trees the leaf is needle-shaped, pointed, and marked by a white line on its under side, Fig. 12(a). In older trees it is scale-like, Fig. 12(b), and the white line on its under side is indistinct.

Range: Widely distributed over nearly all of eastern and central North America.

Soil and location: Grows on poor, gravelly soils as well as in rich bottom lands.

Enemies: The *"cedar apple,"* commonly found on this tree, represents a stage of the apple rust, and for that reason it is not desirable to plant such trees near orchards. Its wood is also sometimes attacked by small *boring insects*.

Value for planting: Its characteristic slender form gives the red cedar an important place as an ornamental tree, but its chief value lies in its commercial use.

Commercial value: The wood is durable, light, smooth and fragrant, and is therefore used for making lead-pencils, cabinets, boxes, moth-proof chests, shingles, posts, and telegraph poles.

Other characters: The *fruit* is small, round and berry-like, about the size of a pea, of dark blue color, and carries from one to four bony seeds.

Other common names: The red cedar is also often called *juniper* and *red juniper*.

Comparisons: The red cedar is apt to be confused with the *low juniper* (*Juniperus communis*) which grows in open fields all over the world. The latter, however, is generally of a low form with a flat top. Its leaves are pointed and prickly, never scale-like, and they are whitish above and green below. Its bark shreds and its fruit is a small round berry of agreeable aromatic odor.

ARBOR-VITAE; NORTHERN WHITE CEDAR (*Thuja occidentalis*)

Distinguishing characters: The **branchlets** are extremely **flat and fan-like**, Fig. 13, and have an agreeable *aromatic odor* when bruised. The tree is an evergreen with a *narrow conical form*.

FIG. 13.—Twig of the Arbor-Vitae.

Leaf: Leaves of two kinds, one scale-like and flat, the other keeled, all tightly pressed to the twig (see Fig. 13).

Form and size: A close, conical head with dense foliage near the base. Usually a small tree, but in some parts of the northeastern States it grows to medium size with a diameter of two feet.

Range: Northern part of North America.

Soil and location: Inhabits low, swampy lands; in the State of Maine often forming thick forests.

Enemies: Very seldom affected by insects.

Value for planting: Is hardy in New England, where it is especially used for hedges. It is also frequently used as a specimen tree on the lawn.

Commercial value: The wood is durable for posts, ties, and shingles. The bark contains considerable tannin and the juices from the tree have a medicinal value.

Other characters: The *fruit* is a cone about ½ inch long.

Other common names: Arbor-vitae is sometimes called *white cedar* and *cedar*.

Comparisons: The arbor-vitae is apt to be confused with the true *white cedar* (*Chamæcyparis thyoides*) but the leaves of the latter are sharp-pointed and not flattened or fan-shaped.

Chapter II

How To Identify Trees—(*Continued*)

GROUP IV. THE LARCH AND CYPRESS

How to tell them from other trees: In summer the larch and cypress may easily be told from other trees by their *leaves*. These are needle-shaped and arranged in clusters with numerous leaves to each cluster in the case of the larch, and feathery and flat in the case of the cypress. In winter, when their leaves have dropped off, the trees can be told by their cones, which adhere to the branches.

There are nine recognized species of larch and two of bald cypress. The larch is characteristically a northern tree, growing in the northern and mountainous regions of the northern hemisphere from the Arctic circle to Pennsylvania in the New World, and in Central Europe, Asia, and Japan in the Old World. It forms large forests in the Alps of Switzerland and France.

The European larch and not the American is the principal species considered here, because it is being planted extensively in this country and in most respects is preferable to the American species.

The bald cypress is a southern tree of ancient origin, the well-known cypress of Montezuma in the gardens of Chepultepec having been a species of Taxodium. The tree is now confined to the swamps and river banks of the South Atlantic and Gulf States, where it often forms extensive forests to the exclusion of all other trees. In those regions along the river swamps, the trees are often submerged for several months of the year.

How to tell them from each other: In summer the larch may be told from the cypress by its leaves (compare Figs. 14 and 16). In winter the two can be distinguished by their characteristic forms. The larch is a broader tree as compared with the cypress and its form is more conical. The cypress is more slender and it is taller. The two have been grouped together in this study because they are both coniferous trees and, unlike the other Conifers, are both deciduous, their leaves falling in October.

FIG. 14.—Twig of the Larch in Summer.

THE EUROPEAN LARCH (*Larix europaea*)

Distinguishing characters: Its leaves, which are needle-shaped and about an inch long, are borne in **clusters** close to the twig, Fig. 14. There are many leaves to each cluster. This characteristic together with the **spire-like** form of the crown will distinguish the tree at a glance.

Leaf: The leaves are of a light-green color but become darker in the spring and in October turn yellow and drop off. The cypress, which is described below, is another cone-bearing tree which sheds its leaves in winter.

FIG. 15.—Twig of the Larch in Winter.

Form and size: A medium-sized tree with a conical head and a straight and tapering trunk. (See Fig. 90.)

Range: Central Europe and eastern and central United States.

Soil and location: Requires a deep, fresh, well-drained soil and needs plenty of light. It flourishes in places where our native species would die. Grows very rapidly.

Enemies: The larch is subject to the attacks of a *sawfly*, which has killed many trees of the American species. A *fungus* (*Trametes pini*) which causes the tree to break down with ease is another of its enemies.

Value for planting: A well-formed tree for the lawn. It is also useful for group planting in the forest.

Commercial value: Because its wood is strong and durable the larch is valuable for poles, posts, railroad ties, and in shipbuilding.

FIG. 16.—Twig of the Cypress.

Other characters: The *fruit* is a small cone about one inch long, adhering to the tree throughout the winter.

FIG. 17.—The Bald Cypress.

Comparisons: The tree is apt to be confused with the *American larch*, also known as *tamarack* and *hackmatack*, but differs from it in having longer leaves, cones twice as large and more abundant and branches which are more pendulous.

The larch differs from the bald cypress in the broader form of its crown and the cluster-like arrangement of its leaves. The twigs of the bald cypress are flat and feathery. The larch and bald cypress have the common characteristics of both shedding their leaves in winter and preferring to grow in moist or swampy soils. The larch, especially the native species, forms the well-known tamarack swamps of the north. The bald cypress grows in a similar way in groups in the southern swamps.

BALD CYPRESS (*Taxodium distichum*)

Distinguishing characters: The **feathery character** of the **twigs**, Fig. 16, and the **spire-like form** of the tree, Fig. 17, which is taller and more slender than the larch, will distinguish this species from others.

FIG. 18.—Cypress "Knees."

Leaf: The leaves drop off in October, though the tree is of the cone-bearing kind. In this respect it is like the larch.

Form and size: Tall and pyramidal.

Range: The cypress is a southern tree, but is found under cultivation in parks and on lawns in northern United States.

Soil and location: Grows naturally in swamps, but will also do well in ordinary well-drained, good soil. In its natural habitat it sends out special roots above water. These are known as "*cypress knees*" (Fig. 18) and serve to provide air to the submerged roots of the tree.

Enemies: None of importance.

Value for planting: An excellent tree for park and lawn planting.

Commercial value: The wood is light, soft, and easily worked. It is used for general construction, interior finish, railroad ties, posts and cooperage.

Other characters: The *bark* is thin and scaly. The *fruit* is a cone about an inch in diameter. The general *color* of the tree is a dull, deep green which, however, turns orange brown in the fall.

Comparisons: The cypress and the larch are apt to be confused, especially in the winter, when the leaves of both have dropped. The cypress is more slender and is taller in form. The leaves of each are very different, as will be seen from the accompanying illustrations.

GROUP V. THE HORSECHESTNUT, ASH AND MAPLE

How to tell them from other trees: The horsechestnut, ash, and maple have their branches and buds arranged on their stems **opposite** each other as shown in Figs. 20, 22 and 24. In other trees, this arrangement is **alternate**, as shown in Fig. 19.

How to tell these three from each other. If the bud is large—an inch to an inch and a half long—dark brown, and *sticky*, it is a *horsechestnut*.

If the bud is *not sticky*, much smaller, and *rusty brown to black* in color, and the ultimate twigs, of an olive green color, are *flattened* at points below the buds, it is an *ash*.

FIG. 19.—Alternate Branching (Beech.)

If it is not a horsechestnut nor an ash and its small buds have many scales covering them, the specimen with branches and buds opposite must then be a *maple*. Each of the maples has one character which distinguishes it from all the other maples. For the sugar maple, this distinguishing character is the *sharp point of the bud*. For the silver maple it is the *bend in the terminal twig*. For the red maple it is the *smooth gray-colored bark*. For the Norway maple it is the *reddish brown color of the full, round bud*, and for the box elder it is the *greenish color of its terminal twig*.

The form of the tree and the leaves are also characteristic in each of the maples, but for the beginner who does not wish to be burdened with too many of these facts at one time, those just enumerated will be found most certain and most easily followed.

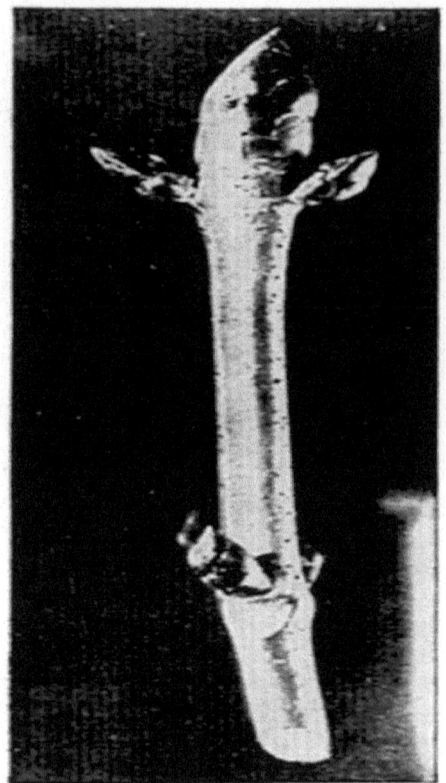

FIG. 20.—Opposite Branching (Horsechestnut.)

THE HORSECHESTNUT (*Æsculus hippocastanum*)

Distinguishing characters: The **sticky** nature of the **terminal bud** and its **large size** (about an inch long). The bud is dark brown in color. See Fig. 20.

Leaf: Five to seven leaflets, usually seven. Fig. 21.

Form and size: Medium-sized tree, pyramidal head and coarse twigs.

Range: Europe and eastern United States.

Soil and location: Prefers a deep, rich soil.

Enemies: The leaves are the favorite food of caterpillars and are subject to a blight which turns them brown prematurely. The trunk is often attacked by a disease which causes the flow of a slimy substance.

Value for planting: On account of its showy flowers, the horsechestnut is a favorite for the park and lawn.

Commercial value: The wood is not durable and is not used commercially.

Other characters: The *flowers* appear in large white clusters in May and June. The *fruit* is large, round, and prickly.

FIG. 21.—Leaf of the Horsechestnut.

Comparisons: The *red horsechestnut* differs from this tree in having red flowers. The *buckeye* is similar to the horsechestnut, but its bud is not sticky and is of a lighter gray color, while the leaf generally has only five leaflets.

THE WHITE ASH (*Fraxinus americana*)

Distinguishing characters: The terminal **twigs** of glossy olive green color are **flattened** below the bud. Fig. 22. The bud is rusty-brown.

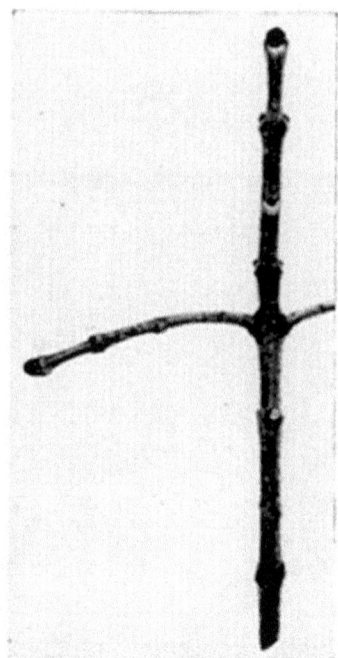

FIG. 22.—Twig of White Ash.

Leaf: Five to nine leaflets. Fig. 23.

Form and size: A large tree with a straight trunk.

Range: Eastern North America.

Soil and location: Rich, moist soil.

Enemies: In cities it is very often attacked by sucking insects.

Value for planting: The white ash grows rapidly. On account of its insect enemies in cities, it should be used more for forest planting and only occasionally for ornament.

Commercial value: It has a heavy, tough, and strong wood, which is valuable in the manufacture of cooperage stock, agricultural implements, and carriages. It is superior in value to the black ash.

Other characters: The *bark* is gray. The *flowers* appear in May.

Comparisons: The white ash is apt to be confused with the *black ash* (*Fraxinus nigra*), but differs from the latter in having a lighter-colored bud. The bud of the black ash is black. The bark of the white ash is darker in color and the terminal twigs are more flattened than those of the black ash.

FIG. 23.—Leaf of White Ash.

SUGAR MAPLE (*Acer saccharum*)

Distinguishing characters: The **bud is sharp-pointed**, scaly, and reddish brown. Fig. 24.

FIG. 24.—Twig of the Sugar Maple.

Leaf: Has sharp points and round sinus. Fig. 25.

Form and size: The crown is oval when the tree is young and round in old age. Fig. 26.

Range: Eastern United States.

Soil and location: Moist and deep soil, and cool, shady positions.

Enemies: Subject to drouth, especially in cities. Is attacked by the *sugar maple borer* and the *maple phenacoccus*, a sucking insect.

Value for planting: Its rich and yellow color in the fall, and the fine spread of its crown make it a desirable tree for the lawn, especially in the country.

Commercial value: Its wood is hard and takes a good polish; used for interior finish and furniture. The tree is also the source of maple sugar. Fig. 27.

Other characters: The *bark* is smooth in young trees and in old trees it shags in large plates. The *flowers* appear in the early part of April.

Other common names: The sugar maple is sometimes called *rock maple* or *hard maple*.

SILVER MAPLE (*Acer saccharinum*)

Distinguishing characters: The tips of the **twigs curve upwards** (Fig. 28), the bark is scaly, and the leaves are very deeply cleft and are silvery on the under side.

FIG. 25.—Leaf of Sugar Maple.

Leaf: Deeply cleft and silvery under side. Fig. 29.

Form and size: A large tree with the main branches separating from the trunk a few feet from the ground. The terminal twigs are long, slender, and drooping.

Range: Eastern United States.

Soil and location: Moist places.

Enemies: The *leopard moth*, a wood-boring insect, and the *cottony-maple scale*, a sucking insect.

FIG. 26.—The Sugar Maple.

Value for planting: Grows too rapidly and is too short-lived to be durable.

Commercial value: Its wood is soft, weak, and little used.

Other characters: The *bark* is light gray, smooth at first and scaly later on. The scales are free at each end and attached in the center. The *flowers* appear before the leaves in the latter part of March or early April.

FIG. 27.—Tapping the Sugar Maple.

Other common names: The silver maple is sometimes known as *soft maple* or *white maple*.

RED MAPLE (*Acer rubrum*)

FIG. 28.—Terminal Twig of Silver Maple.

Distinguishing characters: The **bark is smooth and light gray**, like that of the beech, on the upper branches in older trees, and in young trees over the whole trunk. Fig. 30. The buds are in clusters, and the terminal twigs, Fig. 31, are quite red.

FIG. 29.—Leaf of the Silver Maple.

Leaf: Whitish underneath with three-pointed lobes. Fig. 32.

Form and size: A medium-sized tree with a narrow, round head.

Range: Eastern North America.

Soil and location: Prefers moist places.

Enemies: Leaf blotches (*Rhytisma acerinum*) which, however, are not very injurious.

Value for planting: Suitable as a shade tree for suburban streets. Its rich red leaves in the fall make it attractive for the lawn.

FIG. 30.—Bark of the Red Maple.

Commercial value: Its wood is heavy, close-grained, and takes a good polish. Used for furniture and fuel.

Other characters: The *bud* is small, round, and red. The *flowers* appear before the leaves are out in the early part of April.

FIG. 31.—Twig of the Red Maple.

FIG. 32.—Leaf of the Red Maple.

Other common names: The red maple is sometimes known as *swamp maple*.

FIG. 33.—Twig of Norway Maple.

Comparisons: The red maple is apt to be confused with the silver maple, but the latter can be distinguished by its turned-up twigs and scaly bark over the whole trunk of the tree, which presents a sharp contrast to the straight twig and smooth bark of the red maple. The latter has a bark similar to the beech, but its branches are *opposite*, while those of the beech are *alternate*.

NORWAY MAPLE (*Acer platanoides*)

Distinguishing characters: The bud, Fig. 33, is **oval and reddish-brown** in color; when taken off, a **milky juice exudes**. The bark is close. Fig. 34

FIG. 34.—Bark of Norway Maple.

Leaf: Like the leaf of the sugar maple but thicker in texture and darker in color. Fig. 35.

Form and size: A tall tree with a broad, round head.

Range: Europe and the United States.

Soil and location: Will grow in poor soil.

Enemies: Very few.

Value for planting: One of the best shade trees.

Commercial value: None.

Other characters: The *bark* is close like that of the mockernut hickory.

Comparisons: The Norway maple is apt to be confused with the *sycamore maple* (*Acer pseudoplatanus*), but differs from the latter in having a reddish bud instead of a green bud, and a close bark instead of a scaly bark.

BOX ELDER (*Acer negundo*)

Distinguishing characters: The terminal **twigs are green**, and the buds are round and small. Fig. 36.

Leaf: Has three to seven leaflets.

FIG. 35.—Leaf of Norway Maple.

Form and size: A medium-sized tree with a short trunk and wide-spreading top.

Range: Eastern United States to the Rocky Mountains.

Soil and location: Grows rapidly in deep, moist soil and river valleys, but accommodates itself to the dry and poor soil conditions of the city.

Figure 36.—Twig of Box Elder.

Enemies: Few.

Value for planting: Used as a shade tree in the Middle West, but the tree is so ill formed and so short-lived that it is not to be recommended.

Commercial value: None. The wood is soft.

Other characters: The *bark* of the trunk is smooth and yellowish-green in young trees and grayish brown in older specimens. The *flowers* appear in the early part of April. The *fruit* takes the form of yellowish-green keys which hang on the tree till late fall.

Other common names: The box elder is also commonly known as the *ash-leaf maple*.

GROUP VI. TREES TOLD BY THEIR FORM: ELM, POPLAR, GINGKO AND WILLOW

How to tell them from other trees: The trees described in this group are so distinctive in their general *form* that they may, for the purpose of study, be grouped together, and distinguished from all other trees by this characteristic.

How to tell them from each other: The American elm is *vase-like* in shape; the Lombardy poplar is narrow and *spire-like*; the gingko, or maidenhair tree, is *odd* in its mode of *branching*; and the weeping willow is extremely *pendulous*.

AMERICAN ELM (*Ulmus americana*)

Distinguishing characters: The tree can be told at a glance by its general branching habit. The limbs arch out into a wide-spreading **fan or vase-like crown** which loses itself in numerous fine drooping branchlets. See Fig. 37.

FIG. 37.—American Elm.

Leaf: The leaves are simple, alternate, and from 2 to 5 inches long.

FIG. 38.—English Elm in Winter.

Form and size: It is a tall tree with a trunk that divides a short distance above ground. Its general contour, together with the numerous branches that interlace its massive crown, give the elm an interesting and stately appearance which is unequaled by any other tree.

FIG. 39.—Lombardy Poplar.

Range: Eastern North America.

Soil and location: The elm prefers a deep, rich and moist soil, but will adapt itself even to the poor soil of the city street.

Enemies: *The leopard moth*, a wood-boring insect, and the *elm leaf beetle*, a leaf-eating insect, are the two most important enemies of the tree. Their ravages are very extensive.

Value for planting: The tree has a character of its own which cannot be duplicated for avenue or lawn planting.

Commercial value: The wood is strong and tough and therefore has a special value for cooperage, agricultural implements, carriages, and shipbuilding.

Other characters: The *buds* are small, brown, and smooth, while those of the European elms are covered with down. The *small side twigs* come out at almost right angles to the larger terminal twigs, which is not the case in other species of elm.

FIG. 40.—Leaf of Carolina Poplar.

Other common names: *White elm.*

Comparisons: The *English elm* (*Ulmus campestris*) is also a tall, dignified tree commonly seen under cultivation in America, but may be told from the American species by the difference in their general contour. The branches of the English species spread out but do not arch like those of the American elm, and the bark of the English elm is darker and coarser, Fig. 38. Little tufts of dead twigs along the main branches and trunk of the tree are characteristic of the English elm and will frequently help to distinguish it from the American elm.

The *Camperdown elm* may be recognized readily by its dwarf size and its low drooping umbrella-shaped crown.

LOMBARDY OR ITALIAN POPLAR (*Populus nigra, var. italica*)

Distinguishing characters: Its **tall, slender, spire-like form** and rigidly **erect branches**, which commence low on the trunk, make this tree very distinct at all seasons of the year. See Fig. 39.

Leaf: Triangular in shape, similar to that of the Carolina poplar but smaller, see Fig. 40.

Range: Asia, Europe, and North America.

Soil and location: The poplar is easily grown in poor soil, in any location, and is very hardy.

Value for planting: The tree has a distinctive form which makes it valuable for special landscape effects. It is also used for shelter belts and screening. Like all poplars it is short lived and will stand pruning well.

Commercial value: None.

FIG. 41.—Carolina Poplar.

Comparisons: The *Carolina poplar*, or Cottonwood (*Populus deltoides*) can be told from the Lombardy poplar by its wider crown and its more open branching, Fig. 41. It may be recognized by its big terminal twigs, which are light yellow in color and coarser than those of the Lombardy poplar, Fig. 42. Its bark is smooth, light and yellowish-green in young trees, and dark gray and fissured in older specimens. Its large, conical, glossy, chestnut-brown bud is also characteristic, Fig. 42. Its flowers, in the form of large catkins, a peculiarity of all poplars, appear in the early spring. The Carolina poplar is commonly planted in cities because it grows rapidly and is able to withstand the smoke and drouth conditions of the city. Where other trees, however, can be substituted with success, the poplar should be avoided. Its very fast growth is really a point against the tree, because it grows so fast that it becomes too tall for surrounding property, and its wood being extremely soft and brittle, the tree frequently breaks in windstorms. In many cases it is entirely uprooted, because it is not a deep-rooted tree. Its larger roots, which spread near the surface, upset the sidewalk or prevent the growth of other vegetation on the lawn, while its finer rootlets, in their eager search for moisture, penetrate and clog the joints of neighboring water and sewer pipes. The tree is commonly attacked by the *oyster-shell scale*, an insect which sucks the sap from its bark and which readily spreads to other more valuable trees like the elm. The female form of this tree is even more objectionable than the male, because in the early spring the former produces an abundance of cotton from its seeds which litters the ground and often makes walking dangerous. The only justification for planting the Carolina poplar is in places where the conditions for tree growth are so poor that nothing else will grow, and in those cases the tree should be cut back periodically in order to keep it from becoming too tall and scraggly. It is also desirable for screening in factory districts and similar situations.

FIG. 42.—Bud of the Carolina Poplar.

The *silver* or *white poplar* (*Populus alba*) may be told from the other poplars by its characteristic smooth, *whitish-green bark*, often spotted with dark blotches, Fig. 43. The *leaves are silvery-white* and downy on the under side. The twigs are dark green in color and densely covered with a white down. It grows to very large size and forms an irregular, wide-spreading, broad head, which is characteristically different from that of any of the other poplars.

FIG. 43.—Bark of the Silver Poplar.

The *quaking aspen* (*Populus tremuloides*), the *large-toothed aspen* (*Populus grandidentata*) and the *balsam poplar* or *balm of Gilead* (*Populus balsamifera*) are other common members of the poplar group. The quaking aspen may be told by its reddish-brown twigs, narrow sharp-pointed buds, and by its small finely toothed leaves. The large-toothed aspen has thicker and rather downy buds and broader and more widely toothed leaves. The balsam poplar has a large bud thickly covered with a sticky, pungent, gelatinous substance.

GINGKO OR MAIDENHAIR TREE (*Gingko biloba*)

FIG. 44.—Gingko Trees.

Distinguishing characters: The **peculiar branches** of this tree **emerge upward** from a straight tapering trunk **at an angle of about 45°** and give to the whole tree a striking, Oriental appearance, which is quite different from that of any other tree, Fig. 44.

Leaf: Like that of a leaflet of maidenhair fern, Fig. 45.

Range: A native of northern China and introduced into eastern North America.

Soil and location: The gingko will grow in poor soils.

Enemies: Practically free from insects and disease.

FIG. 45.—Leaves of the Gingko Tree.

Value for planting: It makes a valuable tree for the street where heavy shade is not the object and forms an excellent wide-spreading specimen tree on the lawn.

Other characters: The *fruit* consists of a stone covered by sweet, ill-smelling flesh. The tree is dioecious, there being separate male and female trees. The male tree is preferable for planting in order to avoid the disagreeable odor of the fruit which appears on the female trees when about thirty years old. The male tree has a narrower crown than the female tree. The buds (Fig. 46) are very odd and are conspicuous on the tree throughout the winter. The leaves of the gingko shed in the winter. In this respect the tree is like the larch and the bald cypress.

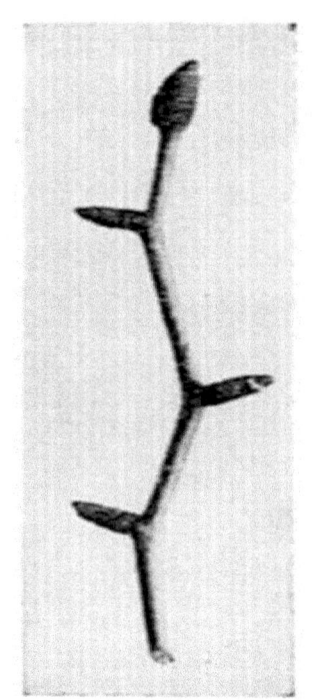

FIG. 46.—Bud of the Gingko Tree.

The gingko belongs to the yew family, which is akin to the pine family. It is therefore a very old tree, the remains of the forests of the ancient world. The gingko in its early life is tall and slender with its few branches close to the stem. But after a time the branches loosen up and form a wide-spreading crown. In the Orient it attains enormous proportions and in this country it also grows to a fairly large size when planted on the open lawn or in groups far apart from other trees so that it can have plenty of room to spread. It then produces a picturesque effect of unusual interest.

WEEPING WILLOW (*Salix babylonica*)

Distinguishing characters: All the willows have a single cap-like scale to the bud, and this species has an unusually **drooping mass of slender branchlets** which characterizes the tree from all others, Fig. 47.

FIG. 47.—Weeping Willow.

Form and size: It grows to large size.

Range: Asia and Europe and naturalized in eastern United States.

Soil and location: Prefers moist places near streams and ponds.

Enemies: None of importance.

Value for planting: The weeping willow has a special ornamental effect in cemeteries and along lakes and river banks in parks.

Commercial value: It is used in the United States for charcoal and for fuel.

Comparisons: The *pussy willow* (*Salix discolor*) may easily be told from the other willows by its small size; it is often no higher than a tall shrub. Its branches are *reddish green* and the buds are dark red, smooth and glossy. The predominating color of the twigs and buds in the pussy willow is therefore a shade of *red*, while in the weeping willow it is *yellowish green*.

GROUP VII. TREES TOLD BY THEIR BARK OR TRUNK: SYCAMORE, BIRCH, BEECH, BLUE BEECH, IRONWOOD, AND HACKBERRY

How to tell them from other trees: The *color of the bark or the form of the trunk* of each of the trees in this group is distinct from that of any other tree.

How to tell them from each other: In the sycamore, the bark is *mottled*; in the white birch, it is *dull white*; in the beech, it is *smooth and gray*; in the hackberry, it is covered with numerous *corky warts*; in the blue beech, the trunk of the tree is *fluted*, as in Fig. 54, and in the ironwood, the bark *peels* in thin perpendicular strips.

FIG. 48.—Bark of the Sycamore Tree.

THE SYCAMORE OR PLANE TREE (*Platanus occidentalis*)

Distinguishing characters: The peculiar **mottled appearance** of the **bark** (Fig. 48) in the trunk and large branches is the striking character here. The bark produces this effect by shedding in large, thin, brittle plates. The newly exposed bark is of a yellowish green color which often turns nearly white later on. **Round seed balls**, about an inch in diameter, may be seen hanging on the tree all winter. In this species, the seed balls are usually solitary, while in the Oriental sycamore, a European tree similar to the native one, they appear in clusters of two, or occasionally of three or four. See Fig. 49.

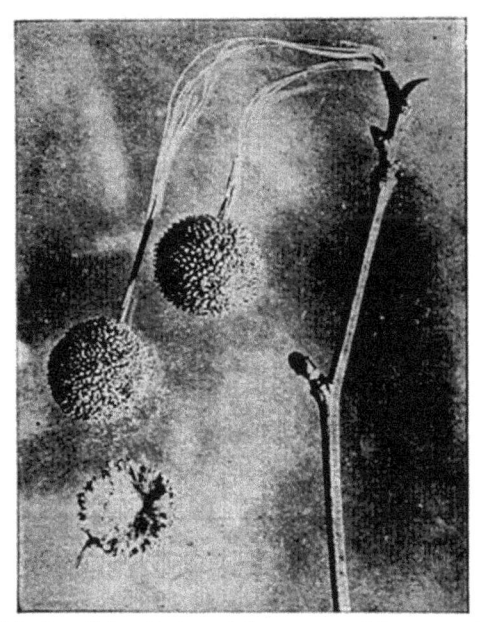

FIG. 49.—Seed-balls of the Oriental Sycamore. Note one Seed-ball cut in half.

FIG. 50.—Gray or White Birch Trees.

Leaf: The stem of the leaf completely covers the bud. This is a characteristic peculiar to sycamores.

Form and size: A large tree with massive trunk and branches and a broad head.

Range: Eastern and southern United States.

Soil and location: Prefers a deep rich soil, but will adapt itself even to the poor soil of the city street.

Enemies: The sycamore is frequently attacked by a fungus (*Gloeosporium nervisequum*), which curls up the young leaves and kills the tips of the branches. Late frosts also often injure its young twigs. The Oriental sycamore, which is the European species, is more hardy in these respects than the native one and is therefore often chosen as a substitute.

Value for planting: The Occidental sycamore is now planted very little, but the Oriental sycamore is used quite extensively in its place, especially as a shade tree. The Oriental sycamore is superior to the native species in many ways. It is more shapely, faster growing, and hardier than the native one. Both sycamores will bear transplanting and pruning well.

FIG. 51.—Bark of the Black or Sweet Birch.

Commercial value: The wood of the sycamore is coarse-grained and hard to work; used occasionally for inside finishing in buildings.

Other names: *Buttonball, buttonwood.*

Comparisons: The *Oriental sycamore* (*Platanus orientalis*) an introduced species, is apt to be confused with the Occidental sycamore, but may be told from the latter by the number of seed balls suspended from the tree. In the case of the Oriental species, the seed balls hang in *pairs* or (rarely) three or four together. In the Occidental, the seed balls are generally *solitary* and very rarely in pairs.

GRAY OR WHITE BIRCH (*Betula populifolia*)

Distinguishing characters: The **dull-white color of the bark** on the trunk and the *dark triangular patches below the insertion of the branches* distinguish this tree; see Fig. 50. The bark of the young trunks and branches is reddish-brown in color and glossy. The bark adheres closely to the trunk of the tree and does not peel in loose, shaggy strips, as in the case of the yellow or golden birch. It is marked by small raised horizontal lines which are the lenticels or breathing pores. These lenticels are characteristic of all birch and cherry trees. In addition to the distinction in the color of the bark, an important character which distinguishes the gray birch from all other species of birch, is found in the **terminal twigs**, which are **rough** to the touch.

Form and size: A small tree. Frequently grows in clumps.

Range: Eastern United States.

Soil and location: The gray birch does best in a deep, rich soil, but will also grow in poor soils.

Enemies: The *bronze-birch borer*, a wood-destroying insect, and *Polyporus betulinus*, a fungus, are its chief enemies.

Value for planting: Its graceful habit and attractive bark gives the tree an important place in ornamental planting. It may be used to advantage with evergreens, and produces a charming effect when planted by itself in clumps.

FIG. 52.—Bark of the Beech.

FIG. 53.—Buds of the Beech Tree.

Commercial value: The wood is soft and not durable. It is used in the manufacture of small articles and for wood pulp.

Other characters: The *fruit is a catkin.*

Comparisons: The *paper birch* (*Betula papyrifera*) is apt to be confused with the gray birch, because both have a white bark. The bark of the paper birch, however, is a clear white and peels off in thin papery layers instead of being close. It very seldom shows any dark triangular markings on the trunk. Its terminal twigs are not rough and its trunk is usually straighter and freer from branches.

The *black* or *sweet birch* (*Betula lenta*) has a bark similar to the gray birch, except that its color is dark gray. See Fig. 51. The twigs have an aromatic taste.

FIG. 54.—Trunk of Blue Beech.

FIG. 55.—Bark of the Ironwood.

The *yellow birch* (*Betula lutea*) has a yellowish or golden bark which constantly peels in thin, ragged, horizontal films.

The *European white birch* (*Betula alba*) has a dull-white bark like the native white birch, but has smooth terminal twigs instead of rough ones. It is commonly seen in the United States on lawns and in parks.

AMERICAN BEECH (*Fagus americana*)

Distinguishing characters: The **close-fitting, smooth, gray bark** will tell this tree from all others except the red maple and yellow-wood. See Fig. 52. The red maple may then be easily eliminated by noting whether the branches are alternate or opposite. They are alternate in the beech and opposite in the maple. The yellow-wood may be eliminated by noting the size of the bud. The **bud** in the yellow-wood is hardly noticeable and of a golden yellow color, while that of the beech is very **long, slender, and sharp-pointed**, and chestnut brown in color. See Fig. 53.

Form and size: It grows tall in the woods, but on the open lawn spreads out into a massive, round-headed tree.

Range: Eastern Canada and United States.

Soil and location: Prefers a rich, well-drained soil, but will grow in any good soil.

Enemies: *Aphides* or plant lice that suck the sap from the leaves in spring and early summer are the chief enemies of the tree.

Value for planting: The pleasing color of its bark, its fine spread of branches, which gracefully droop down to the ground, and its autumnal coloring, make the beech a favorite for lawn and park planting. The several European species of beech are equally charming.

FIG. 56.—Bark of the Hackberry.

Commercial value: The wood is strong, close-grained, and tough. It is used mainly for cooperage, tool handles, shoe lasts, chairs, etc., and for fuel.

Other characters: The *fruit* is a prickly burr encasing a sharply triangular nut which is sweet and edible.

Comparisons: The *European beech* (*Fagus sylvatica*), and its weeping, purple-leaved, and fern-leaved varieties, are frequently met with in parks and may be told from the native species by its darker bark. The weeping form may, of course, be told readily by its drooping branches. The leaves of the European beeches are broader and less serrated than those of the American beech.

BLUE BEECH OR HORNBEAM (*Carpinus caroliniana*)

Distinguishing characters: The **fluted** or muscular effect of its **trunk** will distinguish the tree at a glance, Fig. 54.

Leaf: Doubly serrated; otherwise the same as that of ironwood.

Form and size: A low-spreading tree with branches arching out at various angles, forming a flattened head with a fine, slender spray.

Range: Very common in the eastern United States.

Soil and location: Grows in low wet woods.

Enemies: None of importance.

Value for planting: Its artistic branching and curious trunk give the tree an important place in park planting.

Commercial value: None.

Other characters: The bark is smooth and bluish gray in color.

Comparisons: The blue beech or hornbeam is often confused with the *ironwood* or *hop hornbeam* (*Ostrya virginiana*). The ironwood, however, has a characteristic bark that peels in perpendicular, short, thin segments, often loose at the ends. See Fig. 55. This is entirely different from the close, smooth, and fluted bark of the blue beech. The color of the bark in the ironwood is brownish, while that of the blue beech is bluish-gray. The buds of the ironwood are greenish with brown tips, while the bud of the blue beech shows no green whatever.

HACKBERRY (*Celtis occidentalis*)

Distinguishing characters: The tree may be told readily from other trees by the **corky tubercles** on the bark of the lower portion of the trunk. See Fig. 56.

Leaf: Has three predominating veins and is a bit more developed on one side than on the other.

Form and size: A small or medium-sized tree with a single stem and broad conical crown.

Range: United States and Canada.

Soil and location: Grows naturally in fertile soils, but will adapt itself to almost sterile soils as well.

Enemies: The hackberry is usually free from disease, though often its leaves are covered with insect galls.

Value for planting: It is extensively planted as a shade tree in the Middle West, and is frequently seen as an ornamental tree in the East.

Commercial value: It has little economic value except for fuel.

Other characters: The *fruit* is berry-like, with a hard pit. The fleshy outer part is sweet.

Other common names: *Nettle tree, sugarberry.*

GROUP VIII. THE OAKS AND CHESTNUT

How to tell them from other trees: The oaks are rather difficult to identify and, in studying them it will often be necessary to look for more than one distinguishing character. The oaks differ from other trees in bearing *acorns*. Their *leaves* have many lobes and their upper lateral *buds* cluster at the top of the twigs. The general contour of each oak presents a characteristic branching and sturdiness uncommon in other trees.

The chestnut differs from other trees in bearing *burs* and its *bark* is also distinctly characteristic.

How to tell them from each other: There are two groups of oaks, the *white oak* and the *black oak*. The white oaks mature their acorns in one year and, therefore, only acorns of the same year can be found on trees of this group. The black oaks take two years in which to mature their acorns and, therefore, young acorns of the present year and mature acorns of the previous year may be found on the same tree at one time. The *leaves* of the white oaks have rounded margins and rounded lobes as in Fig. 57, while those of the black oaks have pointed margins and sharp pointed lobes as shown in Figs. 60, 62 and 64. The *bark* of the white oaks is light colored and breaks up in loose flakes as in Fig. 58, while that of the black oaks is darker and deeply ridged or tight as in Figs. 59 and 61. The white oak is the type of the white oak group and the black, red and pin oaks are types of the other. For the characterization of the individual species, the reader is referred to the following pages.

FIG. 57.—Leaf and Fruit of White Oak. (Quercus alba.)

WHITE OAK (*Quercus alba*)

Distinguishing characters: The massive ramification of its branches is characteristic of this species and often an easy clue to its identification. The **bark** has a **light gray color**—lighter than that of the other oaks—and breaks into soft, loose flakes as in Fig. 58. The **leaves are deeply lobed** as in Fig. 57. The **buds are small, round and congested** at the end of the year's growth. The acorns usually have no stalks and are set in shallow, rough cups. The kernels of the acorns are white and palatable.

Form and size: The white oak grows into a large tree with a wide-spreading, massive crown, dissolving into long, heavy, twisted branches. When grown in the open it possesses a short sturdy trunk; in the forest its trunk is tall and stout.

Range: Eastern North America.

FIG. 58.—Bark of White Oak. (Quercus alba.)

Soil and location: The white oak thrives in almost any well-drained, good, deep soil except in a very cold and wet soil. It requires plenty of light and attains great age.

Enemies: The tree is comparatively free from insects and disease except in districts where the Gipsy moth is common, in which case the leaves of the white oak are a favorite food of its caterpillars.

FIG. 59.—Bark of Black Oak. (Quercus velutina).

Value for planting: The white oak is one of the most stately trees. Its massive form and its longevity make the tree suitable for both lawn and woodland planting but it is not used much because it is difficult to transplant and grows rather slowly.

Commercial value: The wood is of great economic importance. It is heavy, hard, strong and durable and is used in cooperage, construction work, interior finish of buildings and for railroad ties, furniture, agricultural implements and fuel.

Comparisons: The *swamp white oak (Quercus platanoides)* is similar to the white oak in general appearance of the bark and form and is therefore liable to be confused with it. It differs from the white oak, however, in possessing a more straggly habit and in the fact that the bark on the under side of its branches shags in loose, large scales. Its buds are smaller, lighter colored and more downy and its acorns are more pointed and with cups more shallow than those of the white oak. The tree also grows in moister ground, generally bordering swamps.

FIG. 60.—Leaf and Fruit of Black Oak. (Quercus velutina).

BLACK OAK (*Quercus velutina*)

Distinguishing characters: The **bark** is black, rough and cut up into firm **ridges** especially at the base of the tree, see Fig. 59. The *inner bark* has a *bright yellow color*. the **leaves** have *sharp points* and are wider at the base than at the tip as shown in Fig. 60. The buds are *large, downy* and *sharp pointed*. The acorns are small and have deep, scaly cups the inner margins of which are downy. The kernels are yellow and bitter.

Form and size: The tree grows in an irregular form to large size, with its branches rather slender as compared with the white oak and with a more open and narrow crown.

Range: Eastern North America.

Soil and location: It will grow in poor soils but does best where the soil is rich and well drained.

Enemies: None of importance.

Value for planting: The black oak is the poorest of the oaks for planting and is rarely offered by nurserymen.

Commercial value: The wood is heavy, hard and strong, but checks readily and is coarse grained. It is of little value except for fuel. The bark is used for tannin.

Other common names: *Yellow oak*.

Comparisons: The black oak might sometimes be confused with the *red* and *scarlet oaks*. The yellow, bitter inner bark will distinguish the black oak from the other two. The light-colored, smooth bark of the red oak and the dark, ridged bark of the black oak will distinguish the two, while the bark of the scarlet oak has an appearance intermediate between the two. The buds of the three species also show marked differences. The buds of the black oak are covered with hairs, those of the scarlet oak have fewer hairs and those of the red are practically free from hairs. The leaves of each of the three species are distinct and the growth habits are different.

RED OAK (*Quercus rubra*)

Distinguishing characters: The **bark** is perpendicularly fissured into long, *smooth, light gray strips* giving the trunk a characteristic **pillar effect** as in Figs. 61 and 94. It has the straightest trunk of all the oaks. The leaves possess *more lobes* than the leaves of any of the other species of the black oak group, see Fig. 62. The acorns, the largest among the oaks, are semispherical with the cups extremely shallow. The buds are large and sharp pointed, but not as large as those of the black oak. They also have a few fine hairs on their scales, but are not nearly as downy as those of the Black oak.

FIG. 61—Bark of Red Oak.

Form and size: The red oak is the largest of the oaks and among the largest of the trees in the northern forests. It has a straight trunk, free from branches to a higher point than in the white oak, see Fig. 94. The branches are less twisted and emerge at sharper angles than do those of the white oak.

Range: It grows all over Eastern North America and reaches north farther than any of the other oaks.

Soil and location: It is less fastidious in its soil and moisture requirements than the other oaks and therefore grows in a great variety of soils. It requires plenty of light.

FIG. 62.—Leaf and Fruit of Red Oak.

Enemies: Like most of the other oaks, this species is comparatively free from insects and disease.

Value for planting: The red oak grows faster and adapts itself better to poor soil conditions than any of the other oaks and is therefore easy to plant and easy to find in the nurseries. It makes an excellent street tree, is equally desirable for the lawn and is hardly surpassed for woodland planting.

Commercial value: The wood is hard and strong but coarse grained, and is used for construction timber, interior finish and furniture. It is inferior to white oak where strength and durability are required.

PIN OAK (*Quercus palustris*)

Distinguishing characters: Its method of **branching** will characterize the tree at a glance. It develops a well-defined *main* ascending *stem* with numerous *drooping* side *branches* as in Fig. 63. The buds are very small and sharp pointed and the leaves are small as in Fig. 64. The bark is dark, firm, smooth and in close ridges. The acorn is small and carries a light brown, striped nut, wider than long and bitter. The cup is shallow, enclosing only the base of the nut.

FIG. 63.—Pin Oaks in Winter.

Form and size: The pin oak is a medium-sized tree in comparison with other oaks. It develops a tall, straight trunk that tapers continuously through a pyramidal crown of low, drooping tender, branches.

Range: Eastern North America.

Soil and location: It requires a deep, rich, moist soil and grows naturally near swamps. Its roots are deep and spreading. The tree grows rapidly and is easily transplanted.

Enemies: None of importance.

Value for planting: The pin oak is an extremely graceful tree and is therefore extensively used for planting on lawns and on certain streets where the tree can find plenty of water and where conditions will permit its branches to droop low.

Commercial value: The wood is heavy and hard but coarse grained and liable to check and warp. Its principal use is in the construction of houses and for shingles.

FIG. 64.—Leaf and Fruit of Pin Oak.

CHESTNUT (*Castanea dentata*)

Distinguishing characters: The **bark** in young trees is smooth and of a marked reddish-bronze color, but when the tree grows older, the bark breaks up into **diamond-shaped ridges**, sufficiently characteristic to distinguish the tree at a glance, see Fig. 65. A close examination of the *terminal twig* will show *three ridges* and *two grooves* running down along the stem from the base of each leaf or leaf-scar. The twig has no true terminal bud. The fruit, a large, round **bur**, prickly without and hairy within and enclosing the familiar dark brown, sweet edible nuts is also a distinguishing mark of the tree.

Leaf: The leaves are distinctly long and narrow. They are from 6 to 8 inches long.

Form and size: The chestnut is a large tree with a massive trunk and broad spreading crown. The chestnut tree when cut, sprouts readily from the stump and therefore in places where the trees have once been cut, a group of two to six trees may be seen emerging from the old stump.

FIG. 65.—Trunk of Chestnut Tree.

Range: Eastern United States.

Soil and location: It will grow on rocky as well as on fertile soils and requires plenty of light.

Enemies: During the past nine years nearly all the chestnut trees in the United States have been attacked by a fungus disease (*Diaporthe parasitica*, Mur.) which still threatens the entire extinction of the chestnut trees in this country. No remedy has been discovered and all affected trees should be cut down and the wood utilized before it decays and becomes worthless. No species of chestnut tree is entirely immune from this disease, though some species are highly resistant.

Value for planting: The chestnut is one of the most rapidly growing hardwood trees but, on account of its disease, which is now prevalent everywhere, it is not wise to plant chestnut trees for the present.

Commercial value: The wood is light, not very strong and liable to warp. It is durable when brought in contact with the soil and is therefore used for railroad ties, fence-posts, poles, and mine timbers. It is also valuable for interior finish in houses and for fuel. Its bark is used in the manufacture of tanning extracts and the nuts are sold in cities in large quantities.

Chapter III

How To Identify Trees—(*Continued*)

GROUP IX. THE HICKORIES, WALNUT AND BUTTERNUT

How to tell them from other trees and from each other: The hickory trees, though symmetrical, have a rugged *appearance* and the *branches* are so sturdy and black as to give a special distinction to this group. The *buds* are different from the buds of all other trees and sufficiently characteristic to distinguish the various species of the group. The *bark* is also a distinguishing character.

The walnut and butternut have *chambered piths* which distinguish them from all other trees and from each other.

SHAGBARK HICKORY (*Hicoria ovata*)

Distinguishing characters: The yellowish brown **buds** nearly as large as those of the mockernut hickory, *are each provided with two long, dark, outer scales* which stand out very conspicuously as shown in Fig. 67. The **bark** in older specimens **shags** off in rough strips, sometimes more than a foot long, as shown in Fig. 68. These two characters will readily distinguish the tree at all seasons of the year.

FIG. 66.—A Shagbark Hickory Tree.

Leaf: The leaf is compound, consisting of 5 or 7 leaflets, the terminal one generally larger.

Form and size: A tall, stately tree—the tallest of the hickories—of rugged form and fine symmetry, see Fig. 66.

Range: Eastern North America.

Soil and location: The shagbark hickory grows in a great variety of soils, but prefers a deep and rather moist soil.

Enemies: The *hickory bark borer (Scolytus quadrispinosus)* is its principal enemy. The insect is now killing thousands of hickory trees in the vicinity of New York City and on several occasions has made its appearance in large numbers in other parts of the country.

Value for planting: It is difficult to transplant, grows slowly and is seldom found in nurseries.

FIG. 67.—Bud of the Shagbark Hickory.

Commercial value: The wood is extremely tough and hard and is used for agricultural implements and for the manufacture of wagons. It is excellent for fuel and the nuts are of great value as a food.

Other characters: The fruit is a nut covered by a thick husk that separates into 4 or 5 segments. The kernel is sweet.

Other common names: *Shellbark hickory.*

MOCKERNUT HICKORY (*Hicoria alba*)

FIG. 68.—Bark of the Shagbark Hickory.

Distinguishing characters: The **bud** is the largest among the hickories—nearly half an inch long—is hard and oval and covered with *yellowish brown* downy *scales* which *do not project* like those of the shagbark hickory, see Fig. 69. The twigs are extremely coarse. The **bark** is very tight on the trunk and branches and has a *close*, hard, *wavy* appearance as in Fig. 70.

Leaf: The leaf consists of 5, 7 or 9 leaflets all of which are large and pubescent and possess a distinct resinous odor.

Form and size: A tall tree with a broad spreading head.

Range: Eastern North America.

Soil and location: The mockernut hickory grows on a great variety of soils, but prefers one which is rich and well-drained.

Enemies: The same as for the shagbark hickory.

Value for planting: It is not commonly planted.

Commercial value: The wood is similar to that of the shagbark hickory and is put to the same uses.

Other characters: The fruit is a nut, larger and covered with a shell thicker than that of the shagbark. The husk is also thicker and separates into four segments nearly to the base. The kernel is small and sweet.

Other common names: *Bigbud hickory; whiteheart hickory.*

Comparisons: The *pignut hickory (Hicoria glabra),* sometimes called broom hickory or brown hickory, often has a shaggy bark, but differs from both the shagbark and the mockernut hickory in possessing buds very much smaller, twigs more slender and leaflets fewer. The nut has a thinner husk which does not separate into four or five segments. The tree prefers drier ground than the other hickories.

FIG. 69.—Bud of the Mockernut Hickory.

The *bitternut (Hicoria minima)* can be told from the mockernut and other species of hickory by its bud, which has no scales at all. The color of its bud is a characteristic orange yellow. The bark is of a lighter shade than the bark of the mockernut hickory and the leaflets are more numerous than in any of the hickories, varying from 7 to 11. Its nuts are bitter.

BLACK WALNUT *(Juglans nigra)*

Distinguishing characters: By cutting a twig lengthwise, it will be seen that its **pith** is divided into little *chambers* as shown in Fig. 71. The bud is dark gray and satiny. The bark is dark brown and deeply ridged and the fruit is the familiar round walnut.

FIG. 70.—Bark of the Mockernut Hickory.

Form and size: A tall tree with a spreading crown composed of stout branches. In the open it grows very symmetrically.

Range: Eastern United States.

Soil and location: The black walnut prefers a deep, rich, fertile soil and requires a great deal of light.

Enemies: The tree is a favorite of many caterpillars.

Value for planting: It forms a beautiful spreading tree on open ground, but is not planted to any extent because it is hard to transplant. It grows slowly unless the soil is very deep and rich, develops its leaves late in the spring and sheds them early in the fall and produces its fruit in great profusion.

Commercial value: The wood is heavy, strong, of chocolate brown color and capable of taking a fine polish. It is used for cabinet making and interior finish of houses. The older the tree, usually, the better the wood, and the consumption of the species in the past has been so heavy that it is becoming rare. The European varieties which are frequently planted in

America as substitutes for the native species yield better nuts, but the American species produces better wood.

FIG. 71.—Twig of the Black Walnut. Note the large chambers in the pith.

FIG. 72.—Twig of the Butternut. Note the small chambers in the pith.

Other characters: The *fruit* is a large round nut about two inches in diameter, covered with a smooth husk which at first is dull green in color and later turns brown. The husk does not separate into sections. The kernel is edible and produces an oil of commercial value.

The *leaves* are compound and alternate with 15 to 23 leaflets to each.

Comparisons: The *butternut (Juglans cinerea)* is another tree that has the pith divided into little chambers, but the little chambers here are shorter than in the black walnut, as may be seen from a comparison of Figs. 71 and 72. The bark of the butternut is light gray while that of the black walnut is dark. The buds in the butternut are longer than those of the black walnut and are light brown instead of gray in color. The form of the tree is low and spreading as compared with the black walnut. The fruit in the butternut is elongated while that of the black walnut is round. The leaves of the butternut have fewer leaflets and these are lighter in color.

GROUP X. TULIP TREE, SWEET GUM, LINDEN, MAGNOLIA, LOCUST, CATALPA, DOGWOOD, MULBERRY AND OSAGE ORANGE

TULIP TREE (*Liriodendron tulipifera*)

Distinguishing characters: There are four characters that stand out conspicuously in the tulip tree—the **bud**, the **trunk**, the persistent **fruit cups** and the wedged **leaf**.

The bud, Fig. 74, about three-quarters of an inch long, is covered by two purplish scales which lend special significance to its whole appearance. The trunk is extremely individual because it rises stout and shaft-like, away above the ground without a branch as shown in Fig. 73. The tree flowers in the latter part of May but the cup that holds the fruit persists throughout the winter. The leaf, Fig. 75, has four lobes, is nearly as broad as it is long and so notched at the upper end that it looks different from any other leaf.

FIG. 73.—The Tulip Tree.

FIG. 74.—Bud of the Tulip Tree.

Form and size: The tulip tree is one of the largest, stateliest and tallest of our trees.

Range: Eastern United States.

Soil and location: Requires a deep, moist soil.

Enemies: Comparatively free from insects and disease.

Value for planting: The tree has great value as a specimen on the lawn but is undesirable as a street tree because it requires considerable moisture and transplants with difficulty. It should be planted while young and where it can obtain plenty of light. It grows rapidly.

Commercial value: The wood is commercially known as *whitewood* and *yellow poplar*. It is light, soft, not strong and easily worked. It is used in construction, for interior finish of houses, woodenware and shingles. It has a medicinal value.

Other characters: The *flower*, shown in Fig. 75, is greenish yellow in color, appears in May and resembles a tulip; hence the name tulip tree. The *fruit* is a cone.

Other common names: *Whitewood; yellow poplar; poplar* and *tulip poplar*.

SWEET GUM (*Liquidambar styraciflua*)

FIG. 75.—Leaf and Flower of the Tulip Tree.

Distinguishing characters: The *persistent, spiny*, long-stemmed round **fruit**; *the corky growths on the* **twigs**, the characteristic *star-shaped* **leaves** (Fig. 76) and the very shiny greenish brown buds and the perfect symmetry of the tree are the chief characters by which to identify the species.

Form and size: The sweet gum has a beautiful symmetrical shape, forming a true monopodium.

FIG. 76.—Leaf and Fruit of the Sweet Gum. Note the corky ridges along the twig.

Range: From Connecticut to Florida and west to Missouri.

Soil and location: Grows in any good soil but prefers low wet ground. It grows rapidly and needs plenty of light.

Enemies: Is very often a favorite of leaf-eating caterpillars.

Value for planting: The tree is sought for the brilliant color of its foliage in the fall, and is suitable for planting both on the lawn and street. In growing the tree for ornamental purposes it is important that it should be frequently transplanted in the nursery and that it be transported with burlap wrapping around its roots.

Commercial value: The wood is reddish brown in color, tends to splinter and is inclined to warp in drying. It is used in cooperage, veneer work and for interior finish.

Other characters: On the smaller branches there are irregular developments of cork as shown in Fig. 76, projecting in some cases to half an inch in thickness.

Other common names: *Red gum.*

Comparisons: The *cork elm* is another tree that possesses corky ridges along its twigs, but this differs from the sweet gum in wanting the spiny fruit and its other distinctive traits.

AMERICAN LINDEN (*Tilia Americana*)

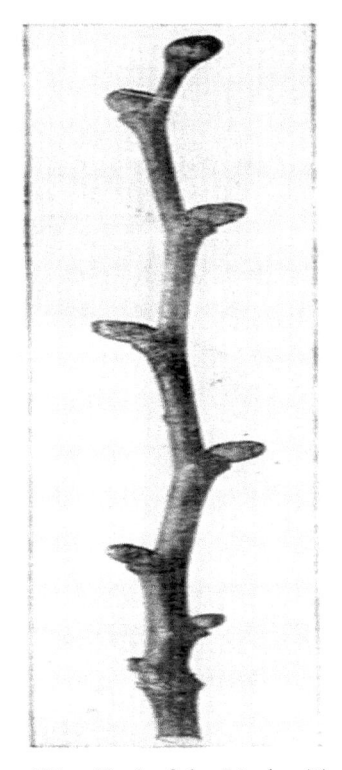

FIG. 77.—Bud of the Linden Tree.

Distinguishing characters: The great distinguishing feature of any linden is the **one-sided** character of its **bud** and **leaf**. The bud, dark red and conical, carries a sort of protuberance which makes it extremely one sided as shown in Fig. 77. The leaf, Fig. 78, is heart-shaped with the side nearest the branch largest.

FIG. 78.—Leaves and Flowers of the European Linden.

Form and size: The American Linden is a medium-sized tree with a broad round head.

Range: Eastern North America and more common in the north than in the south.

Soil and location: Requires a rich, moist soil.

FIG. 79.—European Linden Tree.

FIG. 80.—Bud of the Umbrella Tree.

Enemies: Its leaves are a favorite food of caterpillars and its wood is frequently attacked by a boring insect known as the *linden borer* (*Saperda vestita*).

Value for planting: The linden is easily transplanted and grows rapidly. It is used for lawn and street planting but is less desirable for these purposes than the European species.

Commercial value: The wood is light and soft and used for paper pulp, woodenware, cooperage and furniture. The tree is a favorite with bee keepers on account of the large quantities of nectar contained in its flowers.

Other characters: The *fruit* is like a pea, gray and woody. The *flowers* appear in early July, are greenish-yellow and very fragrant.

Other common names: *Bass-wood; lime-tree; whitewood.*

Comparisons: The *European lindens*, Fig. 79, of which there are several species under cultivation, differ from the native species in having buds and leaves smaller in size, more numerous and darker in color.

THE MAGNOLIAS

The various species of magnolia trees are readily distinguished by their buds. They all prefer moist, rich soil and have their principal value as decorative trees on the lawn. They are distinctly southern trees; some species under cultivation in the United States come from Asia, but the two most commonly grown in the Eastern States are the cucumber tree and the umbrella tree.

FIG. 81.—Bark of the Black Locust.

CUCUMBER TREE (*Magnolia acuminata*)

Distinguishing characters: The **buds** are *small* and *slender* compared with those of the other magnolia trees and are *covered* with small silvery silky *hairs*. The **habit** of the tree is to form a straight axis of great height with a symmetrical mass of branches, producing a perfect monopodial crown. The tree is sometimes known as *mountain magnolia*.

UMBRELLA TREE (*Magnolia tripetala*)

Distinguishing characters: The *buds*, Fig. 80, are extremely *long*, often one and a half inches, have a *purple color* and *are smooth*. The tree does not grow to large size and produces an open spreading head. Its leaves, twelve to eighteen inches long, are larger than those of the other magnolia trees. The tree is sometimes called *elkwood*.

BLACK LOCUST (*Robinia pseudacacia*)

Distinguishing characters: The **bark** of the trunk is *rough* and *deeply ridged*, as shown in Fig. 81. The **buds** are *hardly noticeable*; the twigs sometimes bear small spines on one side. The leaves are large, compound, and fern-like. The individual leaflets are small and delicate.

Form and size: The locust is a medium-sized tree developing a slender straight trunk when grown alongside of others; see Fig. 82.

Range: Canada and United States.

Soil and location: The locust will grow on almost any soil except a wet, heavy one. It requires plenty of light.

Enemies: The *locust borer* has done serious damage to this tree. The grubs of this insect burrow in the sapwood and kill the tree or make it unfit for commercial use. The *locust miner* is a beetle which is now annually defoliating trees of this species in large numbers.

Value for planting: It has little value for ornamental planting.

Commercial value: Though short-lived, the locust grows very rapidly. It is extremely durable in contact with the soil and possesses great strength. It is therefore extensively grown for fence-posts and railroad ties. Locust posts will last from fifteen to twenty years. The wood is valuable for fuel.

FIG. 82.—Black Locust Trees.

Other characters: The *flowers* are showy pea-shaped panicles appearing in May and June. The *fruit* is a small pod.

Other common names: *Yellow locust; common locust; locust.*

Comparisons: The *honey locust* (*Gleditsia triacanthos*) can be told from the black locust by the differences in their bark. In the honey locust the bark is not ridged, has a sort of dark iron-gray color and is often covered with clusters of stout, sharp-pointed thorns as in Fig. 83. The fruit is a large pod often remaining on the tree through the winter. This tree has an ornamental, but no commercial value.

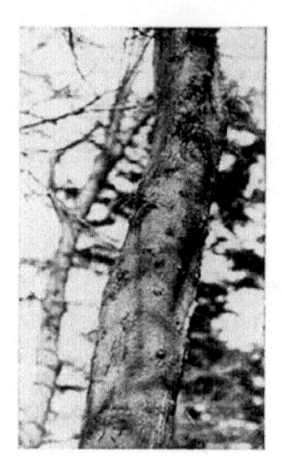

FIG. 83.—Bark of the Honey Locust.

HARDY CATALPA (*Catalpa speciosa*)

Distinguishing characters: The tree may be told by its **fruit**, which hang in long slender pods all winter. The leaf-scars appear on the stem in whorls of three and rarely opposite each other.

Form and size: The catalpa has a short, thick and twisted trunk with an irregular head.

Range: Central and eastern United States.

FIG. 84.—Hardy Catalpa Trees.

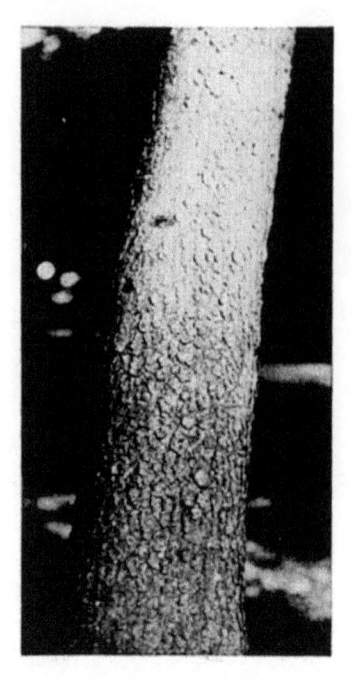

FIG. 85.—Bark of the Flowering Dogwood.

Soil and location: It grows naturally on low bottom-lands but will also do well in poor, dry soils.

Enemies: Practically free from disease and insects.

Value for planting: The catalpa grows very rapidly and is cultivated in parks for ornament and in groves for commercial purposes. The *hardy catalpa* is preferable to the *common catalpa* for planting.

Commercial value: The wood is extremely durable in contact with the soil and is consequently used for posts and railroad ties.

Other characters: The *flowers*, which appear in late June and early July, are large, white and very showy.

Other common names: *Indian bean; western catalpa.*

Comparisons: The *white flowering dogwood* (*Cornus florida*) is a small tree which also has its leaves in whorls of three or sometimes opposite. It can be readily told from other trees, however, by the small square plates into which the outer bark on the trunk divides itself, see Fig. 85, and by the characteristic drooping character of its branches. It is one of the most common plants in our eastern deciduous forests. It is extremely beautiful both in the spring and in the fall and is frequently planted for ornament. There are many varieties of dogwood in common use.

WHITE MULBERRY (*Morus alba*)

A small tree recognized by its *small round reddish brown buds* and *light brown, finely furrowed* (wavy looking) *bark*.

The tree, probably a native of China, is grown under cultivation in eastern Canada and United States. It grows rapidly in moist soil and is not fastidious in its light requirements. Its chief value is for screening and for underplanting in woodlands.

The *red mulberry* (*Morus rubra*) is apt to be confused with the white mulberry, but differs in the following characters: The leaves of the red mulberry are rough on the upper side and downy on the under side, whereas the leaves of the white mulberry are smooth and shiny. The buds in the red are larger and more shiny than those of the white.

The *Osage orange* (*Toxylon pomiferum*) is similar to the mulberry in the light, golden color of its bark, but differs from it in possessing conspicuous spines along the twigs and branches and a more ridged bark.

Chapter IV

The Structure and Requirements of Trees

To be able fully to appreciate trees, their mode of life, their enemies and their care, one must know something of their structure and life requirements.

Structure of trees: Among the lower forms of plants there is very little distinction between the various parts—no differentiation into root, stem, or crown. Often the lower forms of animal and vegetable life are so similar that one cannot discriminate between them. But as we ascend in the scale, the various plant forms become more and more complex until we reach the tree, which is the largest and highest form of all plants. The tree is a living organism composed of cells like any other living organism. It has many parts, every one of which has a definite purpose. The three principal parts are: the stem, the crown, and the root.

The stem: If we examine the cross-section of a tree, Fig. 86, we will notice that it is made up of numerous rings arranged in sections of different color and structure. The central part is known as the *pith*. Around the pith comes a dark, close-grained series of rings known as the *heartwood*, and outside the heartwood comes a lighter layer, the *sapwood*. The *cambium layer* surrounds the sapwood and the *bark* covers all. The cambium layer is the most important tissue of the tree and, together with part of the sapwood, transports the water and food of the tree. It is for this reason that a tree may be hollow, without heart and sapwood, and still produce foliage and fruit.

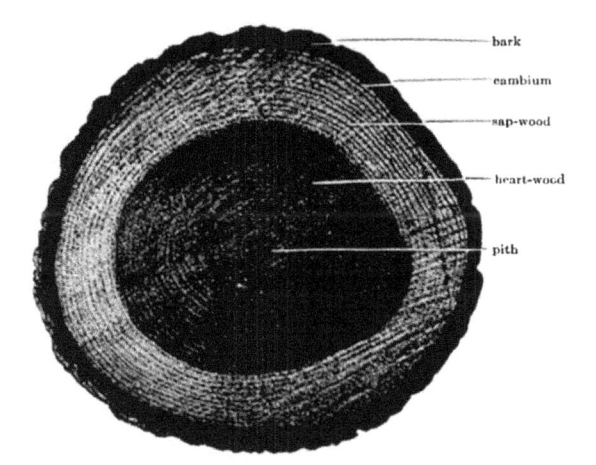

FIG. 86.—The Cross-Section of a Tree.

The crown: The crown varies in form in different species and is developed by the growth of new shoots from buds. The bud grows out to a certain length and forms the branch. Afterwards it thickens only and does not increase in length. New branches will then form from other buds on the same branch. This explains in part the characteristic branching of trees, Fig. 87.

FIG. 87.—Characteristic Form and Branching of Trees. The trees in the photograph are pin oaks.

The leaves are the stomach and lungs of the tree. Their broad blades are a device to catch the sunlight which is needed in the process of digesting the food of the tree. The leaves are arranged on the twigs in such a way as to catch the most sunlight. The leaves take up the carbonic acid gas from the air, decompose it under the influence of light and combine it with the minerals and water brought up by the roots from the soil. The resulting chemical combinations are the sugars and starches used, by the cambium layer in building up the body of the tree. A green pigment, *chlorophyll*, in the leaf is the medium by which, with the aid of sunlight, the sugars are manufactured.

FIG. 88.—Roots of a Hemlock Tree in their Search for Water.

The chlorophyll gives the leaf its green color, and this explains why a tree pales when it is in a dying condition or when its life processes are interfered with. The other colors of the leaf—the reds, browns and yellows of the fall or spring—are due to other pigments. These are angular crystals of different hues, which at certain times of the year become more conspicuous than at others, a phenomenon which explains the variation in the colors of the leaves during the different seasons.

It is evident that a tree is greatly dependent upon its leaves for the manufacture of food and one can, therefore, readily see why it is important to prevent destruction of the leaves by insects or through over-trimming.

The root: The root develops in much the same manner as the crown. Its depth and spread will vary with the species but will also depend somewhat upon the condition of the soil around it. A deep or a dry soil will tend to develop a deep root, while a shallow or moist soil will produce a shallow root, Fig. 88.

The numerous fine hairs which cover the roots serve the purpose of taking up food and water from the soil, while the heavy roots help to support the tree. The root-hairs are extremely tender, are easily dried out when exposed to the sun and wind, and are apt to become overheated when permitted to remain tightly packed for any length of time. These considerations are of practical importance in the planting of trees and in the application of fertilizers. It is these fine rootlets far away from the trunk of the tree that have to be fed, and all fertilizers must, therefore, be applied at points some distance from the trunk and not close to it, where merely the large, supporting roots are located. In the cultivation of trees the same principle holds true.

Requirements of trees: Trees are dependent upon certain soil and atmospheric conditions which influence their growth and development.

(1) *Influence of moisture:* The form of the tree and its growth and structure depend greatly upon the supply of moisture. Botanists have taken the moisture factor as the basis of classification and have subdivided trees into those that grow in moist places (*hydrophytes*), those that grow in medium soils (*mesophytes*), and those that grow in dry places (*xerophytes*). Water is taken up by the roots of the tree from the soil. The liquid absorbed by the roots carries in solution the mineral salts—the food of the tree—and no food can be taken up unless it is in solution. Much of the water is used by the tree and an enormous amount is given off in the process of evaporation.

FIG. 89.—Dead Branches at the Top Caused by Insufficient Water.

These facts will explain some of the fundamental principles in the care of trees. To a tree growing on a city street or on a lawn where nature fails to supply the requisite amount of water, the latter must be supplied artificially, especially during the hot summer months, or else dead branches may result as seen in Fig. 89. Too much thinning out of the crown causes excessive evaporation, and too much cutting out in woodlands causes the soil to dry and the trees to suffer for the want of moisture. This also explains why it is essential, in wooded areas, to retain on the ground the fallen leaves. In decomposing and mixing with the soil, the fallen leaves not only supply the trees with food material, but also tend to conserve moisture in the ground and to prevent the drying out of the soil. Raking off the leaves from wooded areas, a practice common in parks and on private estates—hurts the trees seriously. Some soils may have plenty of moisture, but may also be so heavily saturated with acids or salts that the tree cannot utilize the moisture, and it suffers from drought just the same as if there had been no moisture at all in the soil. Such soils are said to be "physiologically dry" and need treatment.

In the development of disease, moisture is a contributing factor and, therefore, in cavities or underneath bandages where there is likely to be an accumulation of moisture, decay will do more damage than in places that are dry and exposed to the sun.

(2) *Influence of soil:* Soil is made up of fine particles of sand and rock and of vegetable matter called *humus*. A tree will require a certain soil, and unsuitable soils can be very often modified to suit the needs of the tree. A deep, moderately loose, sandy loam, however, which is sufficiently aerated and well supplied with water, will support almost any tree. Too much of any one constituent will make a soil unfit for the production of trees. If too much clay is present the soil becomes "stiff." If too much vegetable matter is present, the soil becomes "sour." The physical character of the soil is also important. By physical character is meant the porosity which results from breaking up the soil. This is accomplished by ploughing or cultivation. In nature, worms help to do this for the soil, but on streets an occasional digging up of the soil about the base of the tree is essential.

Humus or the organic matter in the soil is composed of litter, leaves and animal ingredients that have decayed under the influence of bacteria. The more vegetable matter in the humus, the darker the soil; and therefore a good soil such as one finds on the upper surface of a well-tilled farm has quite a dark color. When, however, a soil contains an unusual quantity of humus, it is known as "muck," and when there is still more humus present we find *peat*. Neither of these two soils is suitable for proper tree growth.

FIG. 90.—A Tree in the Open. Note the full development of the wide crown with branches starting near the ground. The tree is the European larch.

(3) *Influence of light:* Light is required by the leaves in the process of assimilation. Cutting off some of the light from a tree affects its form. This is why trees grown in the open have wide-spreading crowns with branches starting near the ground as in Fig. 90, while the same species growing in the forest produces tall, lanky trees, free from branches to but a few feet from the top as in Fig. 91. Some trees can endure more shade than others, but all will grow in full light. This explains why trees like the beech, hemlock, sugar maple, spruce, holly and dogwood can grow in the shade, while the poplar, birch and willow require light. It also explains why, in the forest, the lower branches die and fall off—a process known in Forestry as "natural pruning," The influence of light on the form of trees should be well understood by all those who plant trees and by those designing landscape effects.

FIG. 91.—A Tree in The Forest. Note the tall stem free from branches and the small, narrow crown.

(4) *Influence of heat:* Trees require a certain amount of heat. They receive it partly from the sun and partly from the soil. Evaporation prevents

the overheating of the crown. The main stem of the tree is heated by water from the soil; therefore trees in the open begin growth in the spring earlier than trees in the forest because the soil in the open is warmer. Shrubs begin their growth earlier than trees because of the nearness of their crowns to their root systems. This also explains why a warm rain will start vegetation quickly. Too much heat will naturally cause excessive drying of the roots or excessive evaporation from the leaves and therefore more water is needed by the tree in summer than in winter.

(5) *Influence of season and frost:* The life processes of a tree are checked when the temperature sinks below a certain point. The tree is thus, during the winter, in a period of rest and only a few chemical changes take place which lead up to the starting of vegetation. In eastern United States, growth starts in April and ceases during the latter part of August or in early September. The different parts of a tree may freeze solid during the winter without injury, provided the tree is a native one. Exotic trees may suffer greatly from extreme cold. This is one of the main reasons why it is always advisable to plant native trees rather than those that are imported and have not yet been acclimatized. Frosts during mid-winter are not quite as injurious as early and late frosts and, therefore, if one is going to protect plants from the winter's cold, it is well to apply the covering early enough and to keep it on late enough to overcome this difficulty.

The mechanical injuries from frost are also important. Snow and sleet will weigh down branches but rarely break them, while frost will cause them to become brittle and to break easily. Those who climb and prune trees should be especially cautious on frosty days.

(6) *Influence of air:* On the under side of leaves and on other surfaces of a tree little pores known as *stomata* may be found. In the bark of birch and cherry trees these openings are very conspicuous and are there known as *lenticels*. These pores are necessary for the breathing of the tree (respiration), whereby carbonic acid gas is taken in from the air and oxygen given out. The process of assimilation depends upon this breathing process and it is therefore evident that when the stomata are clogged as may occur where a tree is subjected to smoke or dust, the life processes of the tree will be interfered with. The same injurious effect results when the stomata of the roots are interfered with. Such interference may occur in cases where a heavy layer of soil is piled around the base of a tree, where the soil about the base of a tree is allowed to become compact, where a tree is planted too deep, or where the roots are submerged under water for any length of time. In any case the air cannot get to the roots and the tree suffers. Nature takes special cognizance of this important requirement in the case of cypress trees, which habitually grow under water. Here the trees are provided with

special woody protuberances known as "cypress knees," which emerge above water and take the necessary air. See Fig. 18.

Conclusions: From the foregoing it will be seen that trees have certain needs that nature or man must supply. These requirements differ with the different species, and in all work of planting and care as well as in the natural distribution of trees it is both interesting and necessary to observe these individual wants, to select species in accordance with local conditions and to care for trees in conformity with their natural needs.

CHAPTER V

WHAT TREES TO PLANT AND HOW

The following classification will show the value of the more important trees for different kinds of planting. The species are arranged in the order of their merit for the particular object under consideration and the comments accompanying each tree are intended to bring out its special qualifications for that purpose.

Conditions for tree growth in one part of the country differ from those of another and these lists, especially applicable to the Eastern States, may not at all fit some other locality.

TREES BEST FOR THE LAWN

DECIDUOUS

1. **American elm** (*Ulmus americana*) — One of the noblest of trees. Possesses a majestic, wide-spreading, umbrella-shaped crown; is easily transplanted, and is suited to a variety of soils.

2. **Pin oak** (*Quercus palustris*) — Has a symmetrical crown with low-drooping branches; requires a moist situation.

3. **European linden** (*Tilia microphylla*) — Possesses a beautiful shade-bearing crown; grows well in ordinary soil.

4. **Red maple** (*Acer rubrum*) — Shows pleasing colors at all seasons; grows best in a fairly rich, moist soil.

5. **Copper beech** (*Fagus sylvatica, alropurpurea*) — Exceedingly beautiful in form, bark, and foliage and possesses great longevity and sturdiness. It is difficult to transplant and therefore only small trees from 6 to 10 feet in height should be used.

6. **Coffee tree** (*Gymnocladus dioicus*) — A unique and interesting effect is produced by its coarse branches and leaves. It is free from insects and disease; requires plenty of light; will grow in poor soils.

7.	**European white birch** (*Belula alba*)	A graceful tree and very effective as a single specimen on the lawn, or in a group among evergreens; should be planted in early spring, and special care taken to protect its tender rootlets.
8.	**Gingko or Maidenhair tree** (*Gingko biloba*)	Where there is plenty of room for the spread of its odd branches, the gingko makes a picturesque specimen tree. It is hardy and free from insect pests and disease.
9.	**Horsechestnut** (*Æsculus hippocastanum*)	Carries beautiful, showy flowers, and has a compact, symmetrical low-branched crown; is frequently subject to insects and disease. The red flowering horsechestnut (*A. rubicunda*) is equally attractive.

FIG. 92.—A Lawn Tree. European Weeping Beech.

10.	**Sugar maple** (*Acer saccharum*)	Has a symmetrical crown and colors beautifully in the fall; requires a rich soil and considerable moisture.
11.	**Soulange's magnolia** (*Magnolia soulangeana*)	Extremely hard and flowers in early spring before the leaves appear.
12.	**Flowering dogwood**	Popular for its beautiful white flowers in the early spring and the rich coloring of its

(Cornus florida)	leaves in the fall; does not grow to large size. The red-flowering variety of this tree, though sometimes not quite as hardy, is extremely beautiful.
13. **Japanese maple** *(Acer polymorphum)*	It has several varieties of different hues and it colors beautifully in the fall; it does not grow to large size.

CONIFEROUS

14. **Oriental spruce** *(Picea orientalis)*	Forms a dignified, large tree with a compact crown and low branches; is hardy.
15. **Austrian pine** *(Pinus austriaca)*	Is very hardy; possesses a compact crown; will grow in soils of medium quality.
16. **Bhotan pine** *(Pinus excelsa)*	Grows luxuriantly; is dignified and beautiful; requires a good soil, and in youth needs some protection from extreme cold.
17. **White pine** *(Pinus strobus)*	Branches gracefully and forms a large, dignified tree; will thrive on a variety of soils.
18. **European larch** *(Larix europaea)*	Has a beautiful appearance; thrives best in moist situations.
19. **Blue spruce** *(Picea pungens)*	Extremely hardy; forms a perfect specimen plant for the lawn.
20. **Japanese umbrella pine** *(Sciadopitys verlicillata)*	Very hardy; retains a compact crown. An excellent specimen plant when grouped with other evergreens on the lawn. Does not grow to large size.
21. **Mugho pine** *(Pinus mughus)*	A low-growing evergreen; hardy; important in group planting.
22. **Obtuse leaf Japanese cypress** *(Retinospora obtusa)*	Beautiful evergreen of small size; hardy; desirable for group planting.
23. **English yew** *(Taxus baccata)*	An excellent evergreen usually of low form; suitable for the lawn, massed with others or as a specimen plant; will grow in the shade of other trees. There are various forms of

this species of distinctive value.

TREES BEST FOR THE STREET

1. **Oriental sycamore**
   (*Platanus orientalis*)

   Very hardy; will adapt itself to city conditions; grows fairly fast and is highly resistant to insects and disease.

2. **Norway maple**
   (*Acer platanoides*)

   Very hardy; possesses a straight trunk and symmetrical crown; is comparatively free from insects and disease and will withstand the average city conditions.

3. **Red oak**
   (*Quercus rubra*)

   Fastest growing of the oaks; very durable and highly resistant to insects and disease; will grow in the average soil of the city street.

FIG. 93.—Street Trees. Norway Maples.

4. **Gingko**
   (*Gingko biloba*)

   Hardy and absolutely free from insects and disease; suited for narrow streets, and will permit of close planting.

5. **European linden**
   (*Tilia microphylla*)

   Beautiful shade-bearing crown; is very responsive to good soil and plenty of moisture.

6. **American elm**
   (*Ulmus americana*)

   When planted in rows along an avenue, it forms a tall majestic archway of great beauty. It is best suited for wide streets and should be planted further apart than the other trees listed above. Requires a fairly good soil and plenty of moisture, and is therefore not suited for planting in the heart of a large city.

7. **Pin oak** (*Quercus palustris*) — This tree exhibits its greatest beauty when its branches are allowed to droop fairly low. It, moreover, needs plenty of moisture to thrive and the tree is therefore best suited for streets in suburban sections, where these conditions can be more readily met.

8. **Red maple** (*Acer rubrum*) — Beautiful in all seasons of the year; requires a rich soil and considerable moisture.

TREES BEST FOR WOODLAND

FOR OPEN PLACES

1. **Red oak** (*Quercus rubra*) — Grows rapidly to large size and produces valuable wood; will grow in poor soil.

2. **White pine** (*Pinus strobus*) — Rapid grower; endures but little shade; wood valuable; will do well on large range of soils.

3. **Red pine** (*Pinus resinosa*) — Very hardy; fairly rapid growing tree.

4. **Tulip tree** (*Liriodendron tulipifera*) — Grows rapidly into a stately forest tree with a clear tall trunk; wood valuable; requires a fairly moist soil. Use a small tree, plant in the spring, and pay special attention to the protection of the roots in planting.

5. **Black locust** (*Robinia pseudacacia*) — Grows rapidly; adapts itself to poor, sandy soils. The wood is suitable for posts and ties.

6. **White ash** (*Fraxinus americana*) — Grows rapidly; prefers moist situations. Wood valuable.

7. **American elm** (*Ulmus americana*) — Grows rapidly to great height; will not endure too much shade; does best in a deep fertile soil. Wood valuable.

8. **European larch**
   (*Larix europaea*)

   Grows rapidly; prefers moist situations.

FIG. 94.—Woodland Trees. Red Oaks.

FOR PLANTING UNDER THE SHADE OF OTHER TREES

9. **Beech**
   (*Fagus*)

   Will stand heavy shade; holds the soil well along banks and steep slopes. Both the American and the English species are desirable.

10. **Hemlock**
    (*Tsuga canadensis*)

    Will stand heavy shade and look effective in winter as well as in summer.

11. **Dogwood**
    (*Cornus florida*)

    Will grow under other trees; flowers beautifully in the spring and colors richly in the fall.

12. **Blue beech**　　　　　Native to the woodlands of the Eastern
　　(*Carpinus caroliniana*)　States; looks well in spring and fall.

TREES BEST FOR SCREENING

1. **Hemlock**　　　　　　Will stand shearing and will screen in winter
　　(*Tsuga canadensis*)　as well as in summer. Plant from 2 to 4 feet
　　　　　　　　　　　　apart to form a hedge.

2. **Osage orange**　　　　Very hardy. Plant close.
　　(*Toxylon pomiferum*)

3. **English hawthorn**　　Flowers beautifully and grows in compact
　　(*Cratægus oxyacantha*)　masses. Plant close.

4. **Lombardy poplar**　　Forms a tall screen and grows under the
　　(*Populus nigra var.*　most unfavorable conditions. Plant 8 to 12
　　italica)　　　　　　　feet apart.

Quality of trees: Trees grown in a nursery are preferable for transplanting to trees grown in the forest. Nursery-grown trees possess a well-developed root system with numerous fibrous rootlets, a straight stem, a symmetrical crown, and a well-defined leader. Trees grown in neighboring nurseries are preferable to those grown at great distances, because they will be better adapted to local climatic and soil conditions. The short distances over which they must be transported also will entail less danger to the roots through drying. For lawn planting, the branches should reach low to the ground, while for street purposes the branches should start at about seven feet from the ground. For street planting, it is also important that the stem should be perfectly straight and about two inches in diameter. For woodland planting, the form of the tree is of minor consideration, though it is well to have the leader well defined here as well as in the other cases. See Fig. 95.

When and how to procure the trees: The trees should be selected in the nursery personally. Some persons prefer to seal the more valuable specimens with leaden seals. Fall is the best time to make the selection, because at that time one can have a wider choice of material. Selecting thus early will also prevent delay in delivery at the time when it is desired to plant.

When to plant: The best time to plant trees is early spring, just before growth begins, and after the frost is out of the ground. From the latter part of March to the early part of May is generally the planting period in the Eastern States.

Where one has to plant both coniferous and deciduous trees, it is best to get the deciduous in first, and then the conifers.

How to plant: The location of the trees with relation to each other should be carefully considered. On the lawn, they should be separated far enough to allow for the full spread of the tree. On streets, trees should be planted thirty to thirty-five feet apart and in case of the elm, forty to fifty feet. In woodlands, it is well to plant as close as six feet apart where small seedlings are used and about twelve feet apart in the case of trees an inch or more in diameter. An abundance of good soil (one to two cubic yards) is essential with each tree where the specimens used are an inch or two in diameter. A rich mellow loam, such as one finds on the surface of a well-tilled farm, is the ideal soil. Manure should never be placed in direct contact with the roots or stem of the tree.

Protection of the roots from drying is the chief precaution to be observed during the planting process, and for this reason a cloudy day is preferable to a sunny day for planting. In case of evergreens, the least exposure of the roots is liable to result disastrously, even more so than in case of deciduous trees. This is why evergreens are lifted from the nursery with a ball of soil around the roots. All bruised roots should be cut off before the tree is planted, and the crown of the tree of the deciduous species should be slightly trimmed in order to equalize the loss of roots by a corresponding decrease in leaf surface.

The tree should be set into the tree hole at the same depth that it stood in the nursery. Its roots, where there is no ball of soil around them, should be carefully spread out and good soil should be worked in carefully with the fingers among the fine rootlets. Every root fibre is thus brought into close contact with the soil. More good soil should be added (in layers) and firmly packed about the roots. The last layer should remain loose so that it may act as a mulch or as an absorbent of moisture. The tree should then be thoroughly watered.

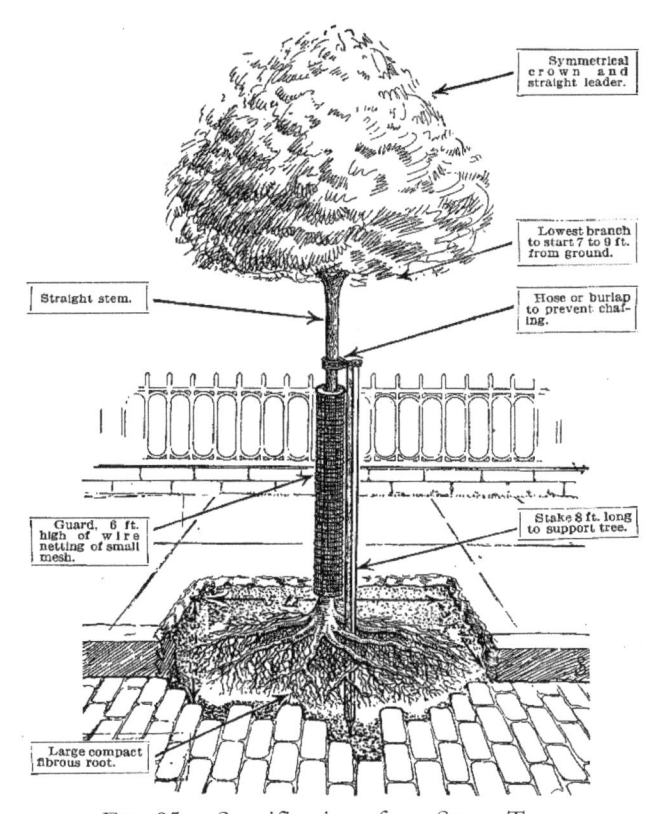

FIG. 95.—Specifications for a Street Tree.

After care: During the first season the tree should be watered and the soil around its base slightly loosened at least once a week, especially on hot summer days. Where trees are planted on streets, near the curb, they should also be fastened to stakes and protected with a wire guard six feet high. See Fig. 95. Wire netting of ½-inch mesh and 17 gauge is the most desirable material.

FIG. 96.—A Home Nursery. (Austrian pines in front.)

Suggestions for a home or school nursery: Schools, farms, and private estates may conveniently start a tree nursery on the premises and raise their own trees. Two-year seedling trees or four-year transplants are best suited for this purpose. These may be obtained from several reliable nurseries in various parts of the country that make a specialty of raising small trees for such purposes. The cost of such trees should be from three to fifteen dollars per thousand.

The little trees, which range from one to two feet in height, will be shipped in bundles. Immediately upon arrival, the bundles should be untied and the trees immersed in a pail containing water mixed with soil. The bundles should then be placed in the ground temporarily, until they can be set out in their proper places. In this process, the individual bundles should be slanted with their tops toward the south, and the spot chosen should be cool and shady. At no time should the roots of these plants be exposed, even for a moment, to sun and wind, and they should always be kept moist. The little trees may remain in this trench for two weeks without injury. They should then be planted out in rows, each row one foot apart for conifers and two feet for broadleaf trees. The individual trees should be set ten inches apart in the row. Careful weeding and watering is the necessary attention later on.

Chapter VI

The Care of Trees

STUDY I. INSECTS INJURIOUS TO TREES AND HOW TO COMBAT THEM

In a general way, trees are attacked by three classes of insects, and the remedy to be employed in each case depends upon the class to which the insect belongs. The three classes of insects are:

1. Those that **chew** and swallow some portion of the leaf; as, for example, the elm leaf beetle, and the tussock, gipsy, and brown-tail moths.

2. Those that **suck** the plant juices from the leaf or bark; such as the San José scale, oyster-shell, and scurfy scales, the cottony maple scale, the maple phenacoccus on the sugar maples, and the various aphides on beech, Norway maple, etc.

3. Those that **bore** inside of the wood or inner bark. The principal members of this class are the leopard moth, the hickory-bark borer, the sugar-maple borer, the elm borer, and the bronze-birch borer.

The chewing insects are destroyed by spraying the leaves with arsenate of lead or Paris green. The insects feed upon the poisoned foliage and thus are themselves poisoned.

The sucking insects are killed by a contact poison: that is, by spraying or washing the affected parts of the tree with a solution which acts externally on the bodies of the insects, smothering or stifling them. The standard solutions for this purpose are kerosene emulsion, soap and water, tobacco extract, or lime-sulfur wash.

FIG. 97.—A Gas-power Spraying Apparatus.

The boring insects are eliminated by cutting out the insect with a knife, by injecting carbon bisulphide into the burrow and clogging the orifice immediately after injection with putty or soap, or in some cases where the tree is hopelessly infested, by cutting down and burning the entire tree.

FIG. 98.—A Barrel Hand-pump Spraying Outfit.

For information regarding the one of these three classes to which any particular insect belongs, and for specific instructions on the application of a remedy, the reader is advised to write to his State Entomologist or to the U. S. Bureau of Entomology at Washington, D. C. The letter should state the name of the tree affected, together with the character of the injury, and should be accompanied by a specimen of the insect, or by a piece of the affected leaf or bark, preferably by both. The advice received will be authentic and will be given without charge.

FIG. 99.—Egg-masses of the Tussock Moth.

When to spray: *In the case of chewing insects*, the latter part of May is the time to spray. The caterpillars hatch from their eggs, and the elm leaf beetle leaves its winter quarters at that time. *In the case of sucking insects*, the instructions will have to be more specific, depending upon the particular insect in question. Some sucking insects can best be handled in May or early June when their young emerge, others can be effectively treated in the fall or winter when the trees are dormant.

How to spray: Thoroughness is the essential principle in all spraying. In the case of leaf-eating insects, this means covering every leaf with the poison and applying it to the under side of the leaves, where the insects generally feed. In the case of sucking insects, thoroughness means an effort to touch every insect with the spray. It should be borne in mind that the insect can be killed only when hit with the chemical. The solution should be well stirred, and should be applied by means of a nozzle that will coat every leaf with a fine, mist-like spray. Mere drenching or too prolonged an application will cause the solution to run off. Special precautions should be taken with contact poisons to see that the formula is correct. Too strong a solution will burn the foliage and tender bark.

Spraying apparatus: There are various forms of spraying apparatus in the market, including small knapsack pumps, barrel hand-pumps, and gasolene and gas-power sprayers, Figs. 97 and 98. Hose and nozzles are essential accessories. One-half inch, three-ply hose of the best quality is necessary to stand the heavy pressure and wear. Two 50-foot lengths is the usual quantity required for use with a barrel hand-pump. Each line of hose should be supplied with a bamboo pole 10 feet long, having a brass tube passed through it to carry the nozzle. The Vermorel nozzle is the best type to use. The cost of a barrel outfit, including two lines of hose, nozzles and truck, should be from $30 to $40. Power sprayers cost from $150 to $300 or more.

Spraying material: *Arsenate of lead* should be used in the proportion Of 4 pounds of the chemical to 50 gallons of water. A brand of arsenate of lead containing at least 14 per cent of arsenic oxide with not more than 50 per cent of water should be insisted upon. This spray may be used successfully against caterpillars and other leaf-eating insects in the spring or summer.

Whale-oil soap should be used at the rate of 1½ pounds of the soap to 1 gallon of hot water, if applied to the tree in winter. As a spray in summer, use 1 pound of the soap to 5 gallons of water. This treatment is useful for most sucking insects.

Lime-sulfur wash is an excellent material to use against sucking insects, such as the San José scale and other armored scales. The application of a lime-sulfur wash when put on during the dormant season is not likely to harm a tree and has such an excellent cleansing effect that the benefits to be derived in this direction alone are often sufficient to meet the cost of the treatment. Lime-sulfur wash consists of a mixture, boiled one hour, of 40 pounds of lime and 80 pounds of sulfur, in 50 gallons of water. It may be had in prepared form and should then be used at the rate of 1 gallon to about 9 gallons of water in winter or early spring before the buds open. At other times of the year and for the softer-bodied insects a more diluted mixture, possibly 1 part to 30 or 40 parts of water, should be used, varying with each case separately.

Kerosene emulsion consists of one-half pound of hard soap, 1 gallon of boiling water, and 2 gallons of kerosene. It may be obtained in prepared form and is then to be used at the rate of one part of the solution to nine parts of water when applied in winter or to the bark only in summer. Use 2 gallons of the solution to a 40-gallon barrel of water when applying it to the leaves in the summer. Kerosene emulsion is useful as a treatment for scale insects.

Tobacco water should be prepared by steeping one-half pound of tobacco stems or leaves in a gallon of boiling water and later diluting the product with 5 to 10 gallons of water. It is particularly useful for plant lice in the summer.

The life history of an insect: In a general way, all insects have four stages of transformation before a new generation is produced. It is important to consider the nature of these four stages in order that the habits of any particular insect and the remedies applicable in combating it may be understood.

All insects develop from *eggs*, Fig. 99. The eggs then hatch into caterpillars or grubs, which is the *larva* stage, in which most insects do the greatest damage to trees. The caterpillars or grubs grow and develop rapidly, and hence their feeding is most ravenous. Following the larva stage comes the third or *pupa* stage, which is the dormant stage of the insect. In this stage the insect curls itself up under the protection of a silken cocoon like the tussock moth, or of a curled leaf like the brown-tail moth, or it may be entirely unsheltered like the pupa of the elm leaf beetle. After the pupa stage comes the *adult insect*, which may be a moth or a beetle.

A study of the four stages of any particular insect is known as a study of its *life history*. The important facts to know about the life history of an insect are the stage in which it does most of its feeding, and the period of the year in which this occurs. It is also important to know how the insect spends the winter in order to decide upon a winter treatment.

IMPORTANT INSECTS
THE ELM LEAF BEETLE

Life history: The elm leaf beetle, Fig. 100, is annually causing the defoliation of thousands of elm trees throughout the United States. Several successive defoliations are liable to kill a tree. The insects pass the winter in the beetle form, hiding themselves in attics and wherever else they can secure shelter. In the middle of May when the buds of the elm trees unfold, the beetles emerge from their winter quarters, mate, and commence eating the leaves, thus producing little holes through them. While this feeding is going on, the females deposit little, bright yellow eggs on the under side of the leaves, which soon hatch into small larvæ or grubs. The grubs then eat away the soft portion of the leaf, causing it to look like lacework. The grubs become full grown in twenty days, crawl down to the base of the tree, and there transform into naked, orange-colored pupæ. This occurs in the early part of August. After remaining in the pupa stage about a week, they change into beetles again, which either begin feeding or go to winter quarters.

Remedies: There are three ways of combating this insect: First, by *spraying the foliage* with arsenate of lead in the latter part of May while the beetles are feeding, and repeating the spraying in June when the larvæ emerge. The spraying method is the one most to be relied on in fighting this insect. A second, though less important remedy, consists in *destroying the pupæ* when they gather in large quantities at the base of the tree. This may be accomplished by gathering them bodily and destroying them, or by pouring hot water or a solution of kerosene over them. In large trees it may be necessary to climb to the crotches of the main limbs to get some of them. The third remedy lies in gathering and *destroying the adult beetles* when found in their winter quarters. The application of bands of burlap or "tanglefoot," or of other substances often seen on the trunks of elm trees is useless, since these bands only prevent the larvæ from crawling down from the leaves to the base and serve to prevent nothing from crawling up. Scraping the trunks of elm trees is also a waste of effort.

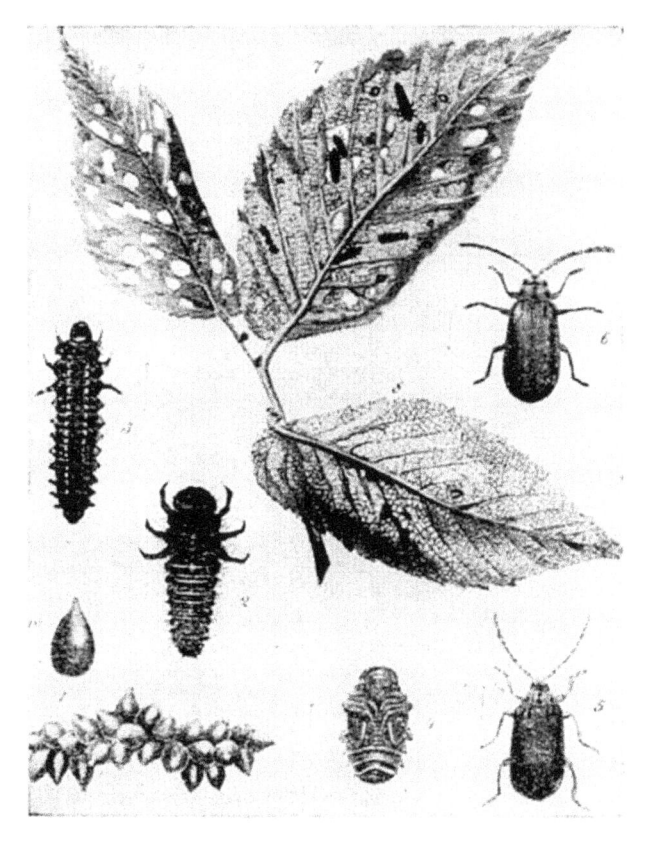

FIG. 100.—The Elm Leaf Beetle. (After Dr. E. P. Felt.)

1. Egg cluster, enlarged. 1a. Single egg, greatly enlarged. 2. Young larva, enlarged. 3. Full grown larva, much enlarged. 4. Pupa, enlarged.
5. Overwintered beetle, enlarged. 6. Fresh, brightly colored beetle, enlarged.
7. Under surface of leaf showing larvæ feeding. 8. Leaf eaten by larvæ.
9. Leaf showing holes eaten by beetles.

THE TUSSOCK MOTH

Life history: This insect appears in the form of a red-headed, yellow-colored caterpillar during the latter part of May, and in June and July. The caterpillars surround themselves with silken cocoons and change into pupæ. The mature moths emerge from the cocoons after a period of about two weeks, and the females, which are wingless, soon deposit their eggs on the bark of trees, on twigs, fences, and other neighboring objects. These eggs form white clusters of nearly 350 individual eggs each, and are very conspicuous all winter, see Fig. 101.

Remedies: There are two ways of combating this insect: (1) By spraying with arsenate of lead for the caterpillars during the latter part of May and early June. (2) By removing and destroying the egg masses in the fall or winter.

FIG. 101.—The Tussock Moth. (After Dr. E. P. Felt.)

1. Caterpillar. 2. Male moth. 3. Female moth laying eggs. 4. Cocoons. 5. Cast skins of caterpillar. 6. Work of young caterpillar. 7. Male pupa. 8 and 9. Girdled branches.

THE GIPSY MOTH

Life history: This insect, imported from Europe to this country in 1868, has ever since proved a serious enemy of most shade, forest, and fruit trees in the New England States. It even feeds on evergreens, killing the trees by a single defoliation.

The insect appears in the caterpillar stage from April to July. It feeds at night and rests by day. The mature caterpillar, which is dark in color, may be recognized by rows of blue and red spots along its back. After July, egg masses are deposited by the female moths on the bark of trees, and on leaves, fences, and other neighboring objects. Here they remain over the winter until they hatch in the spring. The flat egg masses are round or oval in shape, and are yellowish-brown in color. See Fig. 102.

Remedies: Spray for the caterpillars in June with arsenate of lead and apply creosote to the egg masses whenever found.

THE BROWN-TAIL MOTH

Life history: This insect was introduced here from Europe in 1890 and has since done serious damage to shade, forest, and fruit trees, and to shrubs in the New England States.

It appears in the caterpillar stage in the early spring and continues to feed on the leaves and buds until the last of June. Then the caterpillars pupate, the moths come out, and in July and August the egg clusters appear. These hatch into caterpillars which form nests for themselves by drawing the leaves together. Here they remain protected until the spring. See Fig. 103.

Remedies: Collect the winter nests from October to April and burn them. Also spray the trees for caterpillars in early May and especially in August with arsenate of lead.

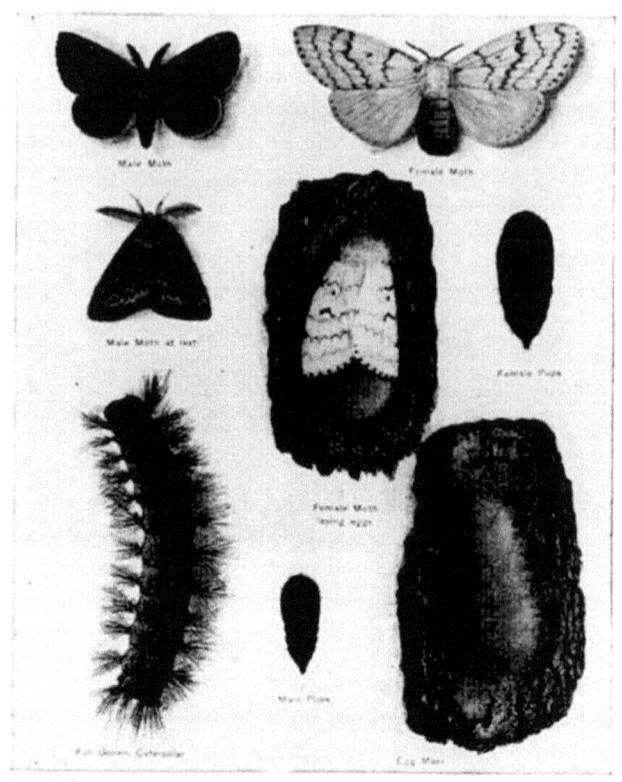

FIG. 102.—The Gipsy Moth. (After F. W. Rane Mass. State Forester.)

FIG. 103.—The Brown-tail Moth. (After F. W. Rane, Mass. State Forester.)

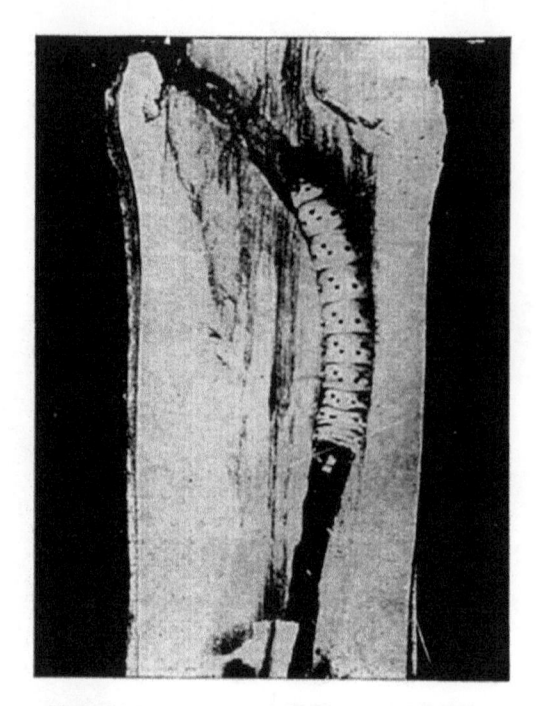

FIG. 104.—Larva of the Leopard Moth.

THE FALL WEBWORM

The caterpillars of this insect congregate in colonies and surround themselves with a web which often reaches the size of a foot or more in diameter. These webs are common on trees in July and August. Cutting off the webs or burning them on the twigs is the most practical remedy.

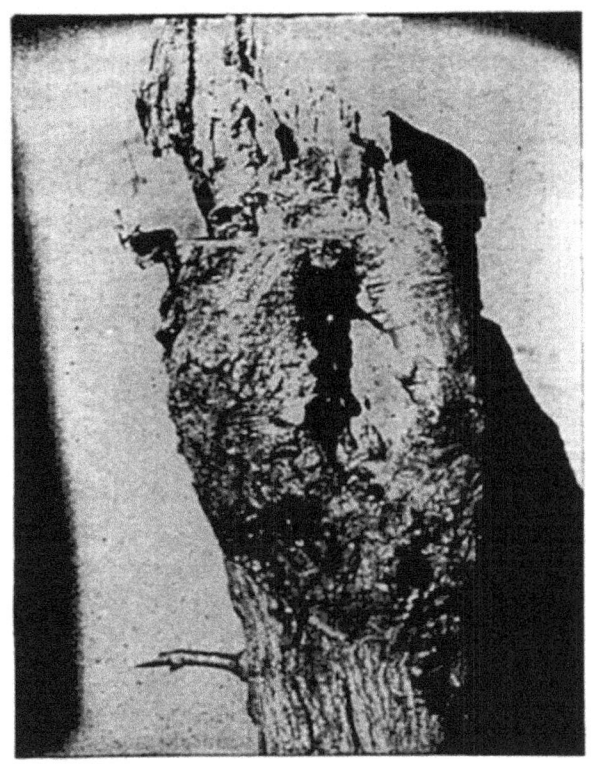

FIG. 105.—Branch Showing Work of the Leopard Moth Larva.

THE LEOPARD MOTH

Life history: This insect does its serious damage in the grub form. The grubs which are whitish in color with brown heads, and which vary in size from 3/8 of an inch to 3 inches in length (Fig. 104), may be found boring in the wood of the branches and trunk of the tree all winter. Fig. 105. The leopard moth requires two years to complete its round of life. The mature moths are marked with dark spots resembling a leopard's skin, hence the name. Fig. 106. It is one of the commonest and most destructive insects in the East and is responsible for the recent death of thousands of the famous elm trees in New Haven and Boston. Fig. 107.

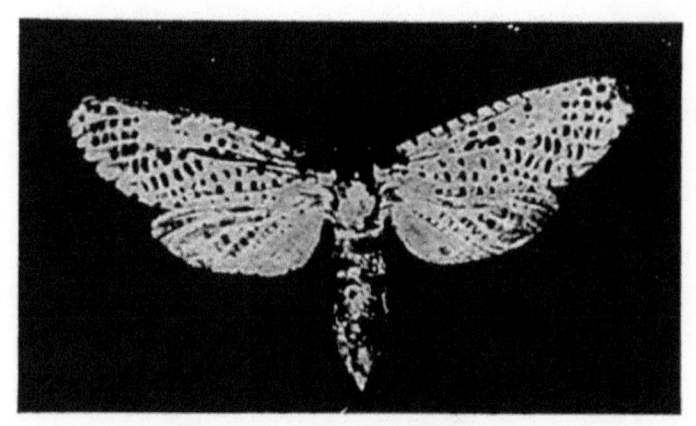

FIG. 106.—The Leopard Moth.

Remedies: Trees likely to be infested with this insect should be examined three or four times a year for wilted twigs, dead branches, and strings of expelled frass; all of which may indicate the presence of this borer. Badly infested branches should be cut off and burned. Trees so badly infested that treatment becomes too complicated should be cut down and destroyed. Where the insects are few and can be readily reached, an injection of carbon bisulphide into the burrow, the orifice of which is then immediately closed with soap or putty, will often destroy the insects within.

FIG. 107.—Elm Tree Attacked by the Leopard Moth.

THE HICKORY BARK BORER

Life history: This insect is a small brown or black beetle in its mature form and a small legless white grub in its winter stage. The beetles appear from June to August. In July they deposit their eggs in the outer sapwood, immediately under the bark of the trunk and larger branches. The eggs soon hatch and the grubs feed on the living tissue of the tree, forming numerous galleries. The grubs pass the winter in a nearly full-grown condition, transform to pupæ in May, and emerge as beetles in June.

Remedies: The presence of the insect can be detected by the small holes in the bark of the trees and the fine sawdust which is ejected from these holes, when the insects are active. It is important to emphasize the advisability of detecting the fine sawdust because that is the best indication of the actual operations of the hickory bark borer. These holes, however, will not be noticeable until the insect has completed its transformation. In summer, the infested trees show wilted leaves and many dead twigs. Holes in the base of the petioles of these leaves are also signs of the working of the insect. Since the insect works underneath the bark, it is inaccessible for treatment and all infested trees should be cut down and burned, or the bark removed and the insects destroyed. This should be done before the beetles emerge from the tree in June.

PLANT LICE OR APHIDES

These often appear on the under side of the leaves of the beech, Norway maple, tulip tree, etc. They excrete a sweet, sticky liquid called "honey-dew," and cause the leaves to curl or drop. Spraying with whale-oil soap solution formed by adding one pound of the soap to five gallons of water is the remedy.

STUDY II. TREE DISEASES

Because trees have wants analogous to those of human beings, they also have diseases similar to those which afflict human beings. In many cases these diseases act like cancerous growths upon the human body; in some instances the ailment may be a general failing due to improper feeding, and in other cases it may be due to interference with the life processes of the tree.

How to tell an ailing tree: Whatever the cause, an ailing tree will manifest its ailment by one or more symptoms.

A change of color in the leaves at a time when they should be perfectly green indicates that the tree is not growing under normal conditions, possibly because of an insufficiency of moisture or light or an overdose of foreign gases or salts. Withering of the leaves is another sign of irregularity in water supply. Dead tops point to some difficulty in the soil conditions or to some disease of the roots or branches. Spotted leaves and mushroom-like growths or brackets protruding from the bark as in Fig. 108, are sure signs of disease.

In attempting to find out whether a tree is healthy or not, one would therefore do well to consider whether the conditions under which it is growing are normal or not; whether the tree is suitable for the location; whether the soil is too dry or too wet; whether the roots are deprived of their necessary water and air by an impenetrable cover of concrete or soil; whether the soil is well drained and free from foreign gases and salts; whether the tree is receiving plenty of light or is too much exposed; and whether it is free from insects and fungi.

If, after a thorough examination, it is found that the ailment has gone too far, it may not be wise to try to save the tree. A timely removal of a tree badly infested with insects or fungi may often be the best procedure and may save many neighboring trees from contagious infection. For this, however, no rules can be laid down and much will depend on the local conditions and the judgment and knowledge of the person concerned.

FIG. 108.—A Bracket Fungus (*Elfvingia megaloma*) on a Tulip Tree.

Fungi as factors of disease: The trees, the shrubs and the flowers with which we are familiar are rooted in the ground and derive their food both from the soil and from the air. There is, however, another group of plants,—*the fungi*,—the roots of which grow in trees and other plants and which obtain their food entirely from the trees or plants upon which they grow. The fungi cannot manufacture their own food as other plants do and consequently absorb the food of their host, eventually reducing it to dust. The fungi are thus disease-producing factors and the source of most of the diseases of trees.

When we can see fungi growing on a tree we may safely assume that they are already in an advanced state of development. We generally discover their presence when their fruiting bodies appear on the surface of the tree as shown in Fig 109. These fruiting bodies are the familiar mushrooms, puffballs, toadstools or shelf-like brackets that one often sees on trees. In some cases they spread over the surface of the wood in thin patches. They vary in size from large bodies to mere pustules barely visible to the naked eye. Their variation in color is also significant, ranging from colorless to black and red but never green. They often emulate the color of the bark, Fig. 110.

Radiating from these fruiting bodies into the tissues of the tree are a large number of minute fibers, comprising the *mycelium* of the fungus. These fibers penetrate the body of the tree in all directions and absorb its food. The mycelium is the most important part of the fungous growth. If the fruiting body is removed, another soon takes its place, but if the entire mycelium is cut out, the fungus will never come back. The fruiting body of the fungus bears the seed or *spores*. These spores are carried by the wind or insects to other trees where they take root in some wound or crevice of the bark and start a new infestation.

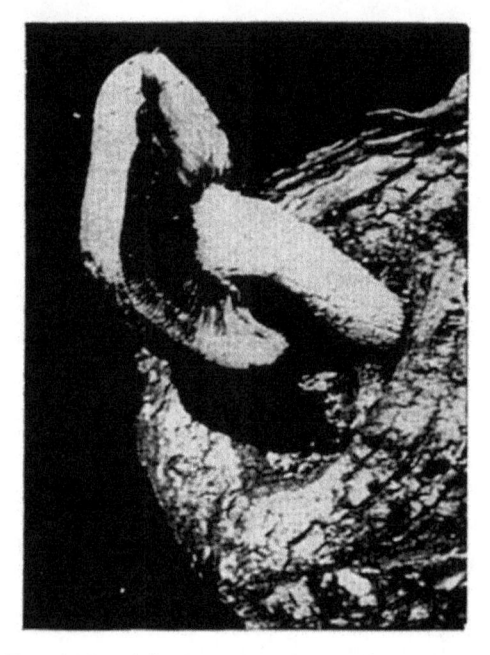

FIG. 109.—The Fruiting Body of a Fungus.

The infestation will be favored in its growth if the spore can find plenty of food, water, warmth and darkness. As these conditions generally exist in wounds and cavities of trees, it is wise to keep all wounds well covered with coal tar and to so drain the cavities that moisture cannot lodge in them. This subject will be gone into more fully in the following two studies on "Pruning Trees" and "Tree Repair."

FIG. 110.—The Birch-fungus rot. (*Polyponis betulinus* Fr.) Note the similarity in the color of the fruiting body and bark of the tree.

While the majority of the fungi grow on the trunks and limbs of trees, some attack the leaves, some the twigs and others the roots. Some fungi grow on living wood some on dead wood and some on both. Those that attack the living trees are the most dangerous from the standpoint of disease.

The chestnut disease: The disease which is threatening the destruction of all the chestnut trees in America is a fungus which has, within recent years, assumed such vast proportions that it deserves special comment. The fungus is known as *Diaporthe parasitica* (Murrill), and was first observed in the vicinity of New York in 1905. At that time only a few trees were known to have been killed by this disease, but now the disease has advanced over the whole chestnut area in the United States, reaching as far south as Virginia and as far west as Buffalo. Fig. 111 shows the result of the chestnut disease.

The fungus attacks the cambium tissue underneath the bark. It enters through a wound in the bark and sends its fungous threads from the point of infection all around the trunk until the latter is girdled and killed. This may all happen within one season. It is not until the tree has practically been destroyed that the disease makes its appearance on the surface of the bark in the form of brown patches studded with little pustules that carry the spores. When once girdled, the tree is killed above the point of infection and everything above dies, while some of the twigs below may live until they are attacked individually by the disease or until the trunk below their origin is infected.

All species of chestnut trees are subject to the disease. The Japanese and Spanish varieties appear to be highly resistant, but are not immune. Other species of trees besides chestnuts are not subject to the disease.

FIG. 111.—Chestnut Trees Killed by the Chestnut Disease.

There is no remedy or preventive for this disease. From the nature of its attack, which is on the inner layer of the tree, it is evident that all applications of fungicides, which must necessarily be applied to the outside of the tree, will not reach the disease. Injections are impossible and other suggested remedies, such as boring holes in the wood for the purpose of inserting chemicals, are futile.

The wood of the chestnut tree, within three or four years after its death, is still sound and may be used for telephone and telegraph poles, posts, railroad ties, lumber and firewood.

Spraying for fungous diseases: Where a fungous disease is attacking the leaves, fruit, or twigs, spraying with Bordeaux mixture may prove effective. The application of Bordeaux mixture is deterrent rather than remedial, and should therefore be made immediately before the disease appears. The nature of the disease and the time of treatment can be determined without cost, by submitting specimens of affected portions of the plant for analysis and advice to the State Agricultural Experiment Station or to the United States Department of Agriculture.

Bordeaux mixture, the standard fungicide material, consists of a solution of 6 pounds of copper sulphate (blue vitriol) with 4 pounds of slaked lime in 50 gallons of water. It may be purchased in prepared form in the open market, and when properly made, has a brilliant sky-blue color. Spraying with Bordeaux mixture should be done in the fall, early spring, or early summer, but never during the period when the trees are in bloom.

STUDY III. PRUNING TREES

FUNDAMENTAL PRINCIPLES

Trees are very much like human beings in their requirements, mode of life and diseases, and the general principles applicable to the care of one are equally important to the intelligent treatment of the other. The removal of limbs from trees, as well as from human beings, must be done sparingly and judiciously. Wounds, in both trees and human beings, must be disinfected and dressed to keep out all fungus or disease germs. Fungous growths of trees are similar to human cancers, both in the manner of their development and the surgical treatment which they require. Improper pruning will invite fungi and insects to the tree, hence the importance of a knowledge of fundamental principles in this branch of tree care.

FIG. 112.—A Tree Pruned Improperly and too Severely.

Time: Too much pruning at one time should never be practiced (Fig. 112), and no branch should be removed from a tree without good reason for so doing. Dead and broken branches should be removed as soon as observed, regardless of any special pruning season, because they are dangerous, unsightly and carry insects and disease into the heart of the tree. But all other pruning, whether it be for the purpose of perfecting the form in shade trees, or for increasing the production of fruit in orchard trees, should be confined to certain seasons. Shade and ornamental trees can best be pruned in the fall, while the leaves are still on the tree and while the tree itself is in practically a dormant state.

Proper cutting: All pruning should be commenced at the top of the tree and finished at the bottom. A shortened branch (excepting in poplars and willows, which should be cut in closely) should terminate in small twigs which may draw the sap to the freshly cut wound; where a branch is removed entirely, the cut should be made close and even with the trunk, as in Fig. 113. Wherever there is a stub left after cutting off a branch, the growing tissue of the tree cannot cover it and the stub eventually decays, falls out and leaves a hole (see Fig. 114), which serves to carry disease and insects to the heart of the tree. This idea of close cutting cannot be over-emphasized.

Where large branches have to be removed, the splitting and ripping of the bark along the trunk is prevented by making one cut beneath the branch, about a foot or two away from the trunk, and then another above, close to the trunk.

FIG. 113.—Branches Properly Cut Close to the Trunk.

Too severe pruning: In pruning trees, many people have a tendency to cut them back so severely as to remove everything but the bare trunk and a few of the main branches. This process is known as "heading back." It is a method, however, which should not be resorted to except in trees that are very old and failing, and even there only with certain species, like the silver maple, sycamore, linden and elm. Trees like the sugar maple will not stand this treatment at all. The willow is a tree that will stand the process very readily and the Carolina poplar must be cut back every few years, in order to keep its crown from becoming too tall, scraggy and unsafe.

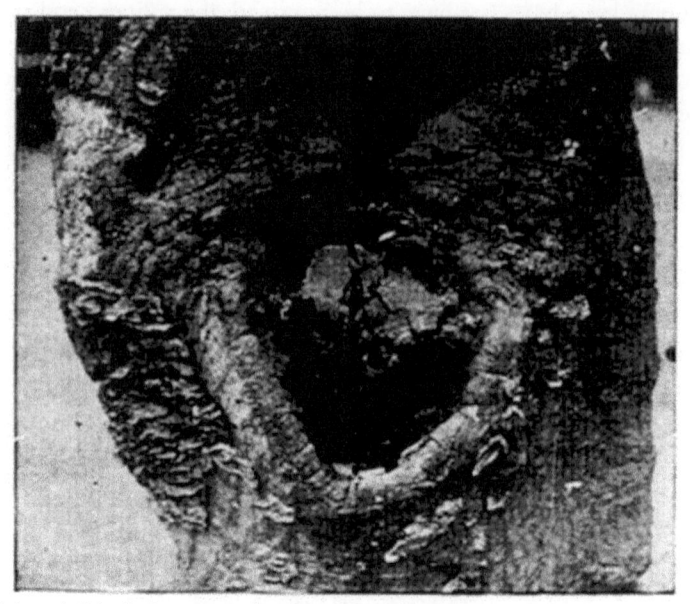

FIG. 114.—A Limb Improperly Cut. Note how the stub is decaying and the resulting cavity is becoming diseased.

Covering wounds: The importance of immediately covering all wounds with coal tar cannot be overstated. If the wound is not tarred, the exposed wood cracks, as in Fig. 115, providing suitable quarters for disease germs that will eventually destroy the body of the tree. Coal tar is by far preferable to paint and other substances for covering the wound. The tar penetrates the exposed wood, producing an antiseptic as well as a protective effect. Paint only forms a covering, which may peel off in course of time and which will later protrude from the cut, thus forming, between the paint and the wood, a suitable breeding place for the development of destructive fungi or disease. The application of tin covers, burlap, or other bandages to the wound is equally futile and in most cases even injurious.

FIG. 115.—Result of a Wound not Covered with Coal Tar. The exposed wood cracked and decay set in.

SPECIAL CONSIDERATIONS

Pruning shade trees: Here, the object is to produce a symmetrical crown and to have the lowest branches raised from the ground sufficiently high to enable pedestrians to pass under with raised umbrellas. Such pruning should, therefore, necessarily be light and confined to the low limbs and dead branches.

Pruning lawn trees: Here the charm of the tree lies in the low reach of the branches and the compactness of the crown. The pruning should, therefore, be limited to the removal of dead and diseased branches only.

Pruning forest trees: Forest trees have a greater commercial value when their straight trunks are free from branches. In the forest, nature generally accomplishes this result and artificial pruning seldom has to be resorted to. Trees in the forest grow so closely together that they shut out the sunlight from their lower limbs, thus causing the latter to die and fall off. This is known as natural pruning. In some European forests, nature is assisted in its pruning by workmen, who saw off the side branches before they fall of their own accord; but in this country such practice would be considered too expensive, hence it is seldom adopted.

TOOLS USED IN PRUNING

Good tools are essential for quick and effective work in pruning. Two or three good saws, a pair of pole-shears, a pole-saw, a 16-foot single ladder, a 40-foot extension ladder of light spruce or pine with hickory rungs, a good pruning knife, plenty of coal tar, a fire-can to heat the tar, a pole-brush, a small hand brush and plenty of good rope comprise the principal equipment of the pruner.

SUGGESTIONS FOR THE SAFETY OF TREE CLIMBERS

1. Before climbing a tree, judge its general condition. The trunk of a tree that shows age, disease, or wood-destroying insects generally has its branches in an equally unhealthy condition.

2. The different kinds of wood naturally differ in their strength and elasticity. The soft and brash woods need greater precautions than the strong and pliable ones. The wood of all the poplars, the ailanthus, the silver maple and the chestnut, catalpa and willow is either too soft or too brittle to be depended upon without special care. The elm, hickory and oak have strong, flexible woods and are, therefore, safer than others. The red oak is weaker than the other oaks. The sycamore and beech have a tough, cross-grained wood which is fairly strong. The linden has a soft wood, while the ash and gum, though strong and flexible, are apt to split.

3. Look out for a limb that shows fungous growths. Every fungus sends fibers into the main body of the limb which draw out its sap. The interior of the branch then loses its strength and becomes like a powder. Outside appearances sometimes do not show the interior condition, but one should regard a fungus as a danger sign.

4. When a limb is full of holes or knots, it generally indicates that borers have been working all kinds of galleries through it, making it unsafe. The silver maple and sycamore maple are especially subject to borers which, in many cases, work on the under side of the branch so that the man in the tree looking down cannot see its dangerous condition.

5. A dead limb with the bark falling off indicates that it died at least three months before and is, therefore, less safe than one with its bark tightly adhering to it.

6. Branches are more apt to snap on a frosty day when they are covered with an icy coating than on a warm summer day.

7. Always use the pole-saw and pole-shears on the tips of long branches, and use the pole-hook in removing dead branches of the ailanthus and other brittle trees where it would be too dangerous to reach them otherwise.

8. Be sure of the strength of a branch before tying an extension ladder to it.

STUDY IV. TREE REPAIR

Where trees have been properly cared for from their early start, wounds and cavities and their subsequent elaborate treatment have no place. But where trees have been neglected or improperly cared for, wounds and cavities are bound to occur and early treatment becomes a necessity.

There are two kinds of wounds on trees: (1) surface wounds, which do not extend beyond the inner bark, and (2) deep wounds or cavities, which may range from a small hole in a crotch to the hollow of an entire trunk.

Surface wounds: Surface wounds (Fig. 116) are due to bruised bark, and a tree thus injured can no longer produce the proper amount of foliage or remain healthy very long. The reason for this becomes very apparent when one looks into the nature of the living or active tissue of a tree and notes how this tissue becomes affected by such injuries.

FIG. 116.—A Surface Wound Properly Freed from Decayed Wood and Covered with Coal Tar.

This living or active tissue is known as the "cambium layer," and is a thin tissue situated immediately under the bark. It must completely envelop the stem, root and branches of the trees. The outer bark is a protective covering to this living layer, while the entire interior wood tissue chiefly serves as a skeleton or support for the tree. The cambium layer is the real, active part of the tree. It is the part which transmits the sap from the base of the tree to its crown; it is the part which causes the tree to grow by the formation of new cells, piled up in the form of rings around the heart of the tree; and it is also the part which prevents the entrance of insects and disease to the inner wood. From this it is quite evident that any injury to the bark, and consequently to this cambium layer alongside of it, will not only cut off a portion of the sap supply and hinder the growth of the tree to an extent proportional to the size of the wound, but will also expose the inner wood to the action of decay. The wound may, at first, appear insignificant, but, if neglected, it will soon commence to decay and thus to carry disease and insects into the tree. The tree then becomes hollow and dangerous and its life is doomed.

Injury to the cambium layer, resulting in surface wounds, may be due to the improper cutting of a branch, to the bite of a horse, to the cut of a knife or the careless wielding of an axe, to the boring of an insect, or to the decay of a fungous disease. (See Fig. 117.) Whatever the cause, *the remedy lies in cleaning out all decayed wood, removing the loose bark and covering the exposed wood with coal tar.*

In cutting off the loose bark, the edges should be made smooth before the coal tar is applied. Loose bark, put back against a tree, will never grow and will only tend to harbor insects and disease. Bandages, too, are hurtful because, underneath the bandage, disease will develop more rapidly than where the wound is exposed to the sun and wind. The application of tin or manure to wounds is often indulged in and is equally injurious to the tree. The secret of all wound treatment is to keep the wound *smooth, clean* to the live tissue, *and well covered* with coal tar.

The chisel or gouge is the best tool to employ in this work. A sharp hawk-billed knife will be useful in cutting off the loose bark. Coal tar is the best material for covering wounds because it has both an antiseptic and a protective effect on the wood tissue. Paint, which is very often used as a substitute for coal tar, is not as effective, because the paint is apt to peel in time, thus allowing moisture and disease to enter the crevice between the paint and the wood.

FIG. 117.—A Neglected Surface Wound. Note the rough surface of the wound, the want of a coal tar covering and the fungous growth that followed.

Cavities: Deep wounds and cavities are generally the result of stubs that have been permitted to rot and fall out. Surface wounds allowed to decay will deepen in course of time and produce cavities. Cavities in trees are especially susceptible to the attack of disease because, in a cavity, there is bound to exist an accumulation of moisture. With this, there is also considerable darkness and protection from wind and cold, and these are all ideal conditions for the development of disease.

The successful application of a remedy, in all cavity treatment, hinges on this principal condition—*that all traces of disease shall be entirely eliminated before treatment is commenced.*

Fungous diseases attacking a cavity produce a mass of fibers, known as the "mycelium," that penetrate the body of the tree or limb on which the cavity is located. In eliminating disease from a cavity, it is, therefore, essential to go *beyond* the mere decaying surface and to cut out all fungous fibers that radiate into the interior of the tree. Where these fibers have penetrated so deeply that it becomes impossible to remove every one of them, the tree or limb thus affected had better be cut down. (Fig. 118.) The presence of the mycelium in wood tissue can readily be told by the discolored and disintegrated appearance of the wood.

The filling in a cavity, moreover, should serve to prevent the accumulation of water and, where a cavity is perpendicular and so located that the water can be drained off without the filling, the latter should be avoided and the cavity should merely be cleaned out and tarred. (Fig. 116.) Where the disease can be entirely eliminated, where the cavity is not too large, and where a filling will serve the practical purpose of preventing the accumulation of moisture, the work of filling should be resorted to.

FIG. 118.—A Cavity Filled in a Tree that Should Have Been Cut Down. Note how the entire interior is decayed and how the tree fell apart soon after treatment.

Filling should be done in the following manner: First, the interior should be thoroughly freed from diseased wood and insects. The chisel, gouge, mall and knife are the tools, and it is better to cut deep and remove every trace of decayed wood than it is to leave a smaller hole in an unhealthy state. The inner surface of the cavity should then be covered with a coat of white lead paint, which acts as a disinfectant and helps to hold the filling. Corrosive sublimate or Bordeaux mixture may be used as a substitute for the white lead paint. A coat of coal tar over the paint is the next step. The cavity is then solidly packed with bricks, stones and mortar as in Fig. 119, and finished with a layer of cement at the mouth of the orifice. This surface layer of cement should not be brought out to the same plane with the outer bark of the tree, but should rather recede a little beyond the growing tissue (cambium layer) which is situated immediately below the bark, Fig. 120. In this way the growing tissue will be enabled to roll over the cement and to cover the whole cavity if it be a small one, or else to grow out sufficiently to overlap the filling and hold it as a frame holds a picture. The cement is used in mixture with sand in the proportion of one-third of cement to two-thirds of sand. When dry, the outer layer of cement should be covered with coal tar to prevent cracking.

FIG. 119.—A Cavity in the Process of being Filled.

FIG. 120—The Same Cavity Properly Filled.

Trees that tend to split: Certain species of trees, like the linden and elm, often tend to split, generally in the crotch of several limbs and sometimes in a fissure along the trunk of the tree. Midwinter is the period when this usually occurs and timely action will save the tree. The remedy lies in fastening together the various parts of the tree by means of bolts or chains.

A very injurious method of accomplishing this end is frequently resorted to, where each of the branches is bound by an iron band and the bands are then joined by a bar. The branches eventually outgrow the diameter of the bands, causing the latter to cut through the bark of the limbs and to destroy them.

Another method of bracing limbs together consists in running a single bolt through them and fastening each end of the bolt with a washer and nut. This method is preferable to the first because it allows for the growth of the limbs in thickness.

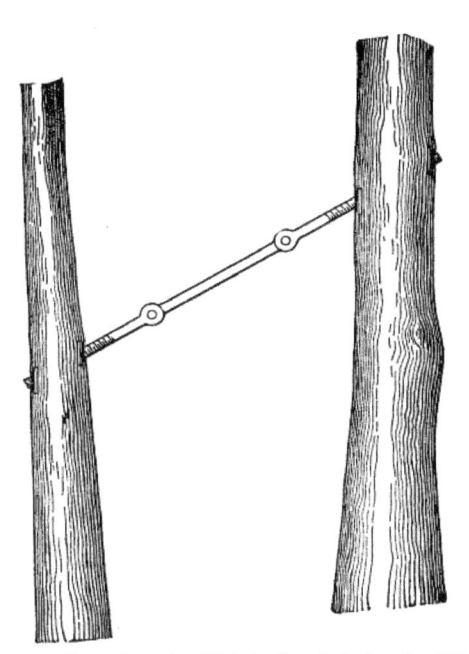

FIG. 121.—Diagram Showing the Triple-bar Method of Fastening Limbs.

A still better method, however, consists in using a bar composed of three parts as shown in Fig. 121. Each of the two branches has a short bolt passed through it horizontally, and the two short bolts are then connected by a third bar. This arrangement will shift all the pressure caused by the swaying of the limbs to the middle connecting-bar. In case of a windstorm, the middle bar will be the one to bend, while the bolts which pass through the limbs will remain intact. The outer ends of the short bolts should have their washers and nuts slightly embedded in the wood of the tree, so that the living tissue of the tree may eventually grow over them in such a way as to hold the bars firmly in place and to exclude moisture and disease. The washers and nuts on the inner side of the limbs should also be embedded.

A chain is sometimes advantageously substituted for the middle section of the bar and, in some cases, where more than two branches have to be joined together, a ring might take the place of the middle bar or chain.

Bolts on a tree detract considerably from its natural beauty and should, therefore, be used only where they are absolutely necessary for the safety of the tree. They should be placed as high up in the tree as possible without weakening the limbs.

Chapter VII

Forestry

STUDY I. WHAT FORESTRY IS AND WHAT IT DOES

Although Forestry is not a new idea but, as a science and an art, has been applied for nearly two thousand years, there are many persons who still need an explanation of its aims and principles.

Forestry deals with the establishment, protection and utilization of forests.

By establishment, is meant the planting of new forests and the cutting of mature forests, in such a way as to encourage a natural growth of new trees without artificial planting or seeding. The planting may consist of sowing seed, or of setting out young trees. The establishment of a forest by cutting may consist of the removal of all mature trees and dependence upon the remaining stumps to reproduce the forest from sprouts, or it may consist of the removal of only a portion of the mature trees, thus giving the young seedlings on the ground room in which to grow.

By protection, is meant the safeguarding of the forest from fire, wind, insects, disease and injury for which man is directly responsible. Here, the forester also prevents injury to the trees from the grazing and browsing of sheep and goats, and keeps his forest so well stocked that no wind can uproot the trees nor can the sun dry up the moist forest soil.

FIG. 122.—A Forest of Bull Pine Cut on Forestry Principles. (Photograph taken on the Black Hills National Forest, South Dakota.)

By utilization, is meant the conservative and intelligent harvesting of the forest, with the aim of obtaining the greatest amount of product from a given area, with the least waste, in the quickest time, and without the slightest deterioration of the forest as a whole. The forester cuts his mature trees, only, and generally leaves a sufficient number on the ground to preserve the forest soil and to cast seed for the production of a new crop. In this way, he secures an annual output without hurting the forest itself. He studies the properties and values of the different woods and places them where they will be most useful. He lays down principles for so harvesting the timber and the by-products of the forest that there will be the least waste and injury to the trees which remain standing. He utilizes the forest, but does not cut enough to interfere with the neighboring watersheds, which the forests protect.

FIG. 123.—A White Pine Plantation, in Rhode Island, Where the Crowns of the Trees Have Met. The trees are fifteen years old and in many cases every other tree had to be removed.

Forestry, therefore, deals with a vast and varied mass of information, comprising all the known facts relating to the life of a forest. It does not deal with the individual tree and its planting and care,—that would be arboriculture. Nor does it consider the grouping of trees for æsthetic effect,—that would be landscape gardening. It concerns itself with the forest as a community of trees and with the utilization of the forest on an economic basis.

Each one of these activities in Forestry is a study in itself and involves considerable detail, of which the reader may obtain a general knowledge in the following pages. For a more complete discussion, the reader is referred to any of the standard books on Forestry.

The life and nature of a forest: When we think of a forest we are apt to think of a large number of individual trees having no special relationship to each other. Closer observation, however, will reveal that the forest consists of a distinct group of trees, sufficiently dense to form an unbroken canopy of tops, and that, where trees grow so closely together, they become very interdependent. It is this interdependence that makes the forest different from a mere group of trees in a park or on a lawn. In this composite character, the forest enriches its own soil from year to year, changes the climate within its own bounds, controls the streams along its borders and supports a multitude of animals and plants peculiar to itself. This communal relationship in the life history of the forest furnishes a most interesting story of struggle and mutual aid. Different trees have different requirements with regard to water, food and light. Some need more water and food than others, some will not endure much shade, and others will grow in the deepest shade. In the open, a tree, if once established, can meet its needs quite readily and, though it has to ward off a number of enemies, insects, disease and windstorm—its struggle for existence is comparatively easy. In the forest, the conditions are different. Here, the tree-enemies have to be battled with, just as in the open, and in addition, instead of there being only a few trees on a plot of ground, there are thousands growing on the same area, all demanding the same things out of a limited supply. The struggle for existence, therefore, becomes keen, many falling behind and but few surviving.

FIG. 124.—Measuring the Diameter of a Tree and Counting its Annual Rings.

This struggle begins with the seed. At first there are thousands of seeds cast upon a given area by the neighboring trees or by the birds and the winds. Of these, only a few germinate; animals feed on some of them, frost nips some and excessive moisture and unfavorable soil conditions prevent others from starting. The few successful ones soon sprout into a number of young trees that grow thriftily until their crowns begin to meet. When the trees have thus met, the struggle is at its height. The side branches encroach upon each other (Fig. 123), shut out the light without which the branches cannot live, and finally kill each other off. The upper branches vie with one another for light, grow unusually fast, and the trees increase in height with special rapidity. This is nature's method of producing clear, straight trunks which are so desirable for poles and large timber. In this struggle for dominance, some survive and tower above the others, but many become stunted and fail to grow, while the majority become entirely overtopped and succumb in the struggle; see Fig. 139.

But in this strife there is also mutual aid. Each tree helps to protect its neighbors against the danger of being uprooted by the wind, and against the sun, which is liable to dry up the rich soil around the roots. This soil is different from the soil on the open lawn. It consists of an accumulation of decayed leaves mixed with inorganic matter, forming, together, a rich composition known as *humus*. The trees also aid each other in forming a close canopy that prevents the rapid evaporation of water from the ground.

The intensity of these conditions will vary a great deal with the composition of the forest and the nature and habits of the individual trees. By composition, or type of forest, is meant the proportion in which the various species of trees are grouped; i.e., whether a certain section of woodland is composed of one species or of a mixture of species. By habit is meant the requirements of the trees for light, water and food.

FIG. 125.—Mountain Slopes in North Carolina Well Covered with Forests.

Some trees will grow in deep shade while others will demand the open. In the matter of water and food, the individual requirements of different trees are equally marked.

The natural rapidity of growth of different species is also important, and one caring for a forest must know this rate of growth, not only as to the individual species, but also with respect to the forest as a whole. If he knows how fast the trees in a forest grow, both in height and diameter, he will know how much wood, in cubic feet, the forest produces in a year, and he can then determine how much he may cut without decreasing the capital stock. The rate of growth is determined in this way: A tree is cut and the rings on the cross-section surface are counted and measured; see Fig. 124. Each ring represents one year's growth. The total number of rings will show the age of the tree. By a study of the rings of the various species of trees on a given plot, the rate of growth of each species in that location can be ascertained and, by knowing the approximate number of trees of each species on the forest area, the rate of growth of the whole forest for any given year can be determined.

FIG. 126.—Bottom Lands Buried in Waste from Deforested Mountains. Wu-t'ai-shan, Shan-si Province, China.

FIG. 127.—Eroded Slope in Western North Carolina.

Forests prevent soil erosion and floods: Forests help to regulate the flow of streams and prevent floods. Most streams are bordered by vast tracts of forest growths. The rain that falls on these forest areas is absorbed and held by the forest soil, which is permeated with decayed leaves, decayed wood and root fibers. The forest floor is, moreover, covered with a heavy undergrowth and thus behaves like a sponge, absorbing the water that falls upon it and then permitting it to ooze out gradually to the valleys and rivers below. A forest soil will retain one-half of its own quantity of water; i.e., for every foot in depth of soil there can be six inches of water and, when thus saturated, the soil will act as a vast, underground reservoir from which the springs and streams are supplied (Fig. 125). Cut the forest down and the land becomes such a desert as is shown in Fig. 126. The soil, leaves, branches and fallen trees dry to dust, are carried off by the wind and, with the fall of rain, the soil begins to wash away and gullies, such as are shown in Fig. 127, are formed. Streams generally have their origins in mountain slopes and there, too, the forests, impeding the sudden run off of the water which is not immediately absorbed, prevent soil erosion.

FIG. 128.—Flood in Pittsburgh, Pa.

Where the soil is allowed to wash off, frequent floods are inevitable. Rain which falls on bare slopes is not caught by the crowns of trees nor held by the forest floor. It does not sink into the ground as readily as in the forest. The result is that a great deal of water reaches the streams in a short time and thus hastens floods. At other periods the streams are low because the water which would have fed them for months has run off in a few days. The farms are the first to suffer from the drouths that follow and, during the period of floods, whole cities are often inundated. Fig. 128 shows such a scene. The history of Forestry is full of horrible incidents of the loss of life and property from floods which are directly traceable to the destruction of the local forests and, on the other hand, there are many cases on record where flood conditions have been entirely obviated by the planting of forests. France and Germany have suffered from inundations resulting from forest devastation and, more than a hundred years ago, both of these countries took steps to reforest their mountain slopes, and thereby to prevent many horrible disasters.

FIG. 129.—Planting a Forest with Seedling Trees on the Nebraska National Forest. The man on the right is placing the tree in a slit just made with the spade. The man on the left is shoveling the dry sand from the surface before making the slit for the tree.

FIG. 130.—Diagrammatic Illustration of a Selection Forest.

How forests are established: New forests may be started from seed or from shoots, or suckers. If from seed, the process may be carried on in one of three ways:

First, by sowing the seed directly on the land.

Second, by first raising young trees in nurseries and later setting them out in their permanent locations in the forest. This method is applicable where quick results are desired, where the area is not too large, or in treeless regions and large open gaps where there is little chance for new trees to spring up from seed furnished by the neighboring trees. It is a method extensively practiced abroad where some of the finest forests are the result. The U. S. government, as well as many of the States, maintain forest-tree nurseries where millions of little trees are grown from seed and planted out on the National and State forests. Fig. 129 shows men engaged in this work. The fundamental principles of starting and maintaining a nursery have already been referred to in the chapter on "What Trees to Plant and How."

The third method of establishing a forest from seed is by cutting the trees in the existing forest so that the seed falling from the remaining trees will, with the addition of light and space, readily take root and fill in the gaps with a vigorous growth of trees, without artificial seeding or planting. This gives rise to several methods of cutting or harvesting forests for the purpose of encouraging natural reproduction. The cutting may extend to single trees over the whole area or over only a part of the whole area. Where the cutting is confined to single trees, the system is known as the "Selection System," because the trees are selected individually, with a view to retaining the best and most vigorous stock and removing the overcrowding specimens and those that are fully mature or infested with disease or insects.

Fig. 130 is a diagrammatic illustration of the operation of this system. In another system the cutting is done in groups, or in strips, and the number of areas of the groups or strips is extended from time to time until the whole forest is cleared. This system is illustrated in Fig. 131. Still another method consists in encouraging trees which will thrive in the shade, such as the beech, spruce and hemlock, to grow under light-demanding trees like the pine. This system presents a "two-storied" forest and is known by that name. The under story often has to be established by planting.

FIG. 131.—Diagrammatic Illustration of the Group or Strip System.

In the system of reproducing forests from shoots or suckers, all trees of a certain species on a given area are cut off and the old stumps and roots are depended upon to produce a new set of sprouts, the strongest of which will later develop into trees. The coniferous trees do not lend themselves at all to this system of treatment, and, among the broadleaf trees, the species vary in their ability to sprout. Some, like the chestnut and poplar, sprout profusely; others sprout very little.

How forests are protected: Forestry also tries to protect the forests from many destructive agencies. Wasteful lumbering and fire are the worst enemies of the forest. Fungi, insects, grazing, wind, snow and floods are the other enemies.

FIG. 132.—The Result of a Forest Fire. The trees, lodgepole pine and Englemann spruce, are all dead and down. Photograph taken in the Colorado National Forest, Colorado.

By wasteful lumbering is meant that the forest is cut with no regard for the future and with considerable waste in the utilization of the product. Conservative lumbering, which is the term used by foresters to designate the opposite of wasteful lumbering, will be described more fully later in this study.

Protection from fire is no less important than protection from wasteful lumbering. Forest fires are very common in this country and cause incalculable destruction to life and property; see Fig. 132. From ten to twelve million acres of forest-land are burnt over annually and the timber destroyed is estimated at fifty millions of dollars. The history of Forestry abounds in tales of destructive fires, where thousands of persons have been killed or left destitute, whole towns wiped out, and millions of dollars in property destroyed. In most cases, these uncontrollable fires started from small conflagrations that could readily, with proper fire-patrol, have been put out.

There are various ways of fighting fires, depending on the character of the fire,—whether it is a surface fire, burning along the surface layer of dry leaves and small ground vegetation, a ground fire, burning below the surface, through the layer of soil and vegetable matter that generally lines the forest floor, or a top fire, burning high up in the trees.

When the fire runs along the surface only, the injury extends to the butts of the trees and to the young seedlings. Such fires can be put out by throwing dirt or sand over the fire, by beating it, and, sometimes, by merely raking the leaves away.

Ground fires destroy the vegetable mold which the trees need for their sustenance. They progress slowly and kill or weaken the roots of the trees.

FIG. 133.—A Top Fire near Bear Canyon, Arizona.

Top fires, Fig. 133, are the most dangerous, destroying everything in their way. They generally develop from surface fires, though sometimes they are started by lightning. They are more common in coniferous forests, because the leaves of hardwoods do not burn so readily. Checking the progress of a top fire is a difficult matter. Some fires will travel as rapidly as five miles an hour, and the heat is terrific. The only salvation for the forest lies, in many cases, in a sudden downpour of rain, a change of wind, or some barrier which the fire cannot pass. A barrier of this kind is often made by starting another fire some distance ahead of the principal one, so that when the two fires meet, they will die out for want of fuel. In well-kept forests, strips or lanes, free from inflammable material, are often purposely made through the forest area to furnish protection against top fires. Carefully managed forests are also patrolled during the dry season so that fires may be detected and attacked in their first stages. Look-out stations, watch-towers, telephone-connections and signal stations are other means frequently resorted to for fire protection and control. Notices warning campers and trespassers against starting fires are commonly posted in such forests. (Fig. 143.)

FIG. 134.—Sheep Grazing on Holy Cross National Forest, Colorado. The drove consists of 1600 sheep, of which only part are shown in the photograph.

The grazing of sheep, goats and cattle in the forest is another important source of injury to which foresters must give attention. In the West this is quite a problem, for, when many thousands of these animals pass through a forest (Fig. 134), there is often very little young growth left and the future reproduction of the forest is severely retarded. Grazing on our National Forests is regulated by the Government.

As a means of protection against insects and fungi, all trees infested are removed as soon as observed and in advance of all others, whenever a lumbering operation is undertaken.

FIG. 135.—A Typical Montana Sawmill.

How forests are harvested: Forestry and forest preservation require that a forest should be cut and not merely held untouched. But it also demands that the cutting shall be done on scientific principles, and that only as much timber shall be removed in a given time as the forest can produce in a corresponding period. After the cutting, the forest must be left in a condition to produce another crop of timber within a reasonable time: see Fig. 122. These fundamental requirements represent the difference between conservative lumbering and ordinary lumbering. Besides insuring a future supply of timber, conservative lumbering, or lumbering on forestry principles, also tends to preserve the forest floor and the young trees growing on it, and to prevent injury to the remaining trees through fire, insects and disease. It provides for a working plan by which the kind, number and location of the trees to be cut are specified, the height of the stumps is stipulated and the utilization of the wood and by-products is regulated.

Conservative lumbering provides that the trees shall be cut as near to the ground as possible and that they shall be felled with the least damage to the young trees growing near by. The branches of the trees, after they have been felled, must be cut and piled in heaps, as shown in Fig. 122, to prevent fire. When the trunks, sawed into logs, are dragged through the woods, care is taken not to break down the young trees or to injure the bark of standing trees. Waste in the process of manufacture is provided against, uses are found for the material ordinarily rejected, and the best methods of handling and drying lumber are employed. Fig. 135 shows a typical sawmill capable of providing lumber in large quantities.

In the utilization of the by-products of the forest, such as turpentine and resin, Forestry has devised numerous methods for harvesting the crops with greater economy and with least waste and injury to the trees from which the by-products are obtained. Fig. 136 illustrates an improved method by which crude turpentine is obtained.

FIG. 136.—Gathering Crude Turpentine by the Cup and Gutter Method. This system, devised by foresters, saves the trees and increases the output.

Forestry here and abroad: Forestry is practiced in every civilized country except China and Turkey. In Germany, Forestry has attained, through a long series of years, a remarkable state of scientific thoroughness and has greatly increased the annual output of the forests of that country.

In France, Switzerland, Austria, Hungary, Norway, Sweden, Russia and Denmark, Forestry is also practiced on scientific principles and the government in each of these countries holds large tracts of forests in reserve. In British India one finds a highly efficient Forest Service and in Japan Forestry is receiving considerable attention.

In the United States, the forest areas are controlled by private interests, by the Government and by the States. On privately owned forests,

Forestry is practiced only in isolated cases. The States are taking hold of the problem very actively and in many of them we now find special Forestry Commissions authorized to care for vast areas of forest land reserved for State control. These Commissions employ technically trained foresters who not only protect the State forests, but also plant new areas, encourage forest planting on private lands and disseminate forestry information among the citizens. New York State has such a Commission that cares for more than a million acres of forest land located in the northern part of the State. Many other States are equally progressive.

The United States Government is the most active factor in the preservation of our forests. The Government to-day owns over two hundred million acres of forest land, set aside as National Forests. There are one hundred and fifty individual reserves, distributed as shown in Fig. 137 and cared for by the Forest Service, a bureau in the Department of Agriculture. Each of the forests is in charge of a supervisor. He has with him a professional forester and a body of men who patrol the tract against fire and the illegal cutting of timber. Some of the men are engaged in planting trees on the open areas and others in studying the important forest problems of the region. Fig. 138.

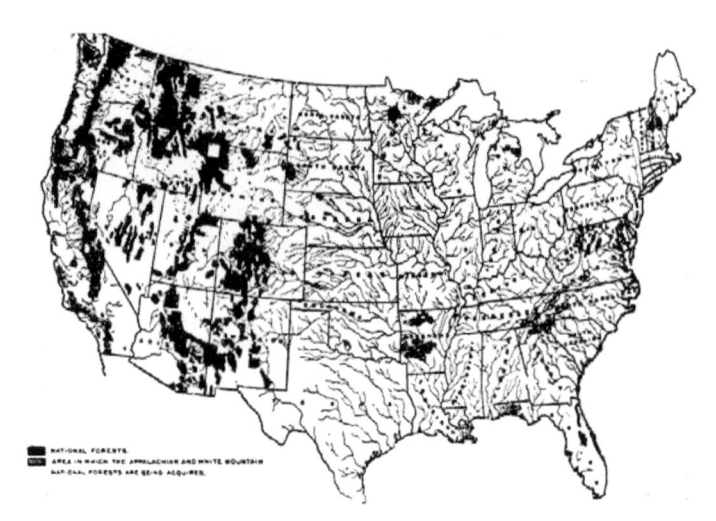

FIG. 137.—Map Showing Our National Forests. (larger version)

FIG. 138.—Government Foresters in Missouri Studying the Growth and Habits of Trees. They are standing in water three feet deep.

Where cutting is to be done on a National Forest, the conditions are investigated by a technically trained forester and the cutting is regulated according to his findings. Special attention is given to discovering new uses for species of trees which have hitherto been considered valueless, and the demand upon certain rare species is lessened by introducing more common woods which are suitable for use in their place.

Aside from the perpetuation of the national forests, the U. S. Forest Service also undertakes such tree studies as lie beyond the power or means of private individuals. It thus stands ready to cooperate with all who need assistance.

STUDY II. CARE OF THE WOODLAND

Almost every farm, large private estate or park has a wooded area for the purpose of supplying fuel or for enhancing the landscape effect of the place. In most instances these wooded areas are entirely neglected or are so improperly cared for as to cause injury rather than good. In but very few cases is provision made for a future growth of trees after the present stock has gone. Proper attention will increase and perpetuate a crop of good trees just as it will any other crop on the farm, while the attractiveness of the place may be greatly enhanced through the intelligent planting and care of trees.

How to judge the conditions: A close examination of the wooded area may reveal some or all of the following unfavorable conditions:

The trees may be so crowded that none can grow well. A few may have grown to large size but the rest usually are decrepit, and overtopped by the larger trees. They are, therefore, unable, for the want of light and space, to develop into good trees. Fig. 139 shows woodland in such condition.

FIG. 139.—Woodland which Needs Attention. The trees are overcrowded.

There may also be dead and dying trees, trees infested with injurious insects and fungi and having any number of decayed branches. The trees may be growing so far apart that their trunks will be covered with suckers as far down as the ground, or there may be large, open gaps with no trees at all. Here the sun, striking with full force, may be drying up the soil and preventing the decomposition of the leaves. Grass soon starts to grow in these open spaces and the whole character of the woodland changes as shown in Figs. 140 and 141.

FIG. 140.—First Stage of Deterioration. The woodland is too open and grass has taken the place of the humus cover.

Where any of these conditions exist, the woodland requires immediate attention. Otherwise, as time goes on, it deteriorates more and more, the struggle for space among the crowded and suppressed trees becomes more keen, the insects in the dying trees multiply and disease spreads from tree to tree. Under such conditions, the soil deteriorates and the older trees begin to suffer.

FIG. 141.—Second Stage of Deterioration. The Surface Soil of the Wooded Area Has Washed Away and the Trees Have Died.

The attention required for the proper care of woodland may be summed up under the four general heads of *soil preservation, planting, cutting,* and *protection.*

Improvement by soil preservation: The soil in a wooded area can best be preserved and kept rich by doing two things; by retaining the fallen leaves on the ground and by keeping the ground well covered with a heavy growth of trees, shrubs and herbaceous plants. The fallen leaves decompose, mix with the soil and form a dark-colored material known as *humus.* The humus supplies the tree with a considerable portion of its food and helps to absorb and retain the moisture in the soil upon which the tree is greatly dependent. A heavy growth of trees and shrubs has a similar effect by serving to retain the moisture in the soil.

Improvement by planting: The planting of new trees is a necessity on almost any wooded area. For even where the existing trees are in good condition, they cannot last forever, and provision must be made for others to take their place after they are gone. The majority of the wooded areas in our parks and on private estates are not provided with a sufficient undergrowth of desirable trees to take the place of the older ones. Thus, also, the open gaps must be planted to prevent the soil from deteriorating.

Waste lands on farms which are unsuited for farm crops often offer areas on which trees may profitably be planted. These lands are sufficiently good in most cases to grow trees, thus affording a means of turning into value ground which would otherwise be worthless. It has been demonstrated that the returns from such plantations at the end of fifty years will yield a six per cent investment and an extra profit of $151.97 per acre, the expense totaling at the end of fifty years, $307.03. The value of the land is estimated at $4 per acre and the cost of the trees and planting at $7 per acre. The species figured on here is white pine, one of the best trees to plant from a commercial standpoint. With other trees, the returns will vary accordingly.

FIG. 142.—A Farm Woodlot.

The usual idea that it costs a great deal to plant several thousand young trees is erroneous. An ordinary woodlot may be stocked with a well-selected number of young trees at a cost less than the price generally paid for a dozen good specimen trees for the front lawn. It is not necessary to underplant the woodlot with big trees. The existing big trees are there to give character to the forest and the new planting should be done principally as a future investment and as a means of perpetuating the life of the woodlot. Young trees are even more desirable for such planting than the older and more expensive ones. The young trees will adapt themselves to the local soil and climatic conditions more easily than the older ones. Their demand for food and moisture is more easily satisfied, and because of their small cost, one can even afford to lose a large percentage of them after planting.

The young plants should be two-year-old seedlings or three-year-old "transplants."

Two-year-old seedlings are trees that have been grown from the seed in seed beds until they reach that age. They run from two to fifteen inches in height, depending upon the species.

Three-year-old "transplants" have been grown from the seed in seed beds and at the end of the first or second year have been taken up and transplanted into rows, where they grow a year or two longer. They are usually a little taller than the two-year-old seedlings, are much stockier and have a better root system. For this reason, three-year-old transplants are a little more desirable as stock for planting. They will withstand drought better than seedlings.

The best results from woodland planting are obtained with native-grown material. Such stock is stronger, hardier and better acclimated. Foreign-grown stock is usually a little cheaper, owing to the fact that it has been grown abroad, under cheap labor conditions.

The trees may be purchased from reputable dealers, of whom there are many in this country. These dealers specialize in growing young trees and selling them at the low cost of three to ten dollars per thousand. In States in which a Forestry Commission has been inaugurated, there have also been established State nurseries where millions of little trees are grown for reforestation purposes. In order to encourage private tree planting, the Forestry Commissions are usually willing to sell some of these trees at cost price, under certain conditions, to private land owners. Inquiries should be made to the State Forestry Commission.

Great care must be taken to select the species most suitable for the particular soil, climatic and light conditions of the woodlot. The trees which are native to the locality and are found growing thriftily on the woodlot, are the ones that have proven their adaptability to the local conditions and should therefore be the principal species used for underplanting. A list from which to select the main stock would, therefore, vary with the locality. In the Eastern States it would comprise the usual hardy trees like the red, pin and scarlet oaks, the beech, the red and sugar maples, the white ash, the tulip tree, sycamore, sweet gum and locust among the deciduous trees; the white, Austrian, red, pitch and Scotch pines, the hemlock and the yew among the conifers.

With the main stock well selected, one may add a number of trees and shrubs that will give to the woodland scene a pleasing appearance at all seasons. The brilliant autumnal tints of the sassafras, pepperidge, blue beech, viburnum, juneberry and sumach are strikingly attractive. The flowering dogwood along the drives and paths will add a charm in June as well as in autumn and an occasional group of white birch will have the same effect if planted among groups of evergreens. Additional undergrowth of native woodland shrubs, such as New Jersey tea, red-berried elder and blueberry for the Eastern States, will augment the naturalness of the scene and help to conserve the moisture in the soil.

Two or three years' growth will raise these plants above all grass and low vegetation, and a sprinkling of laurel, rhododendron, hardy ferns and a few intermingling colonies of native wild flowers such as bloodroot, false Solomon's seal and columbines for the East, as a ground cover will put the finishing touches to the forest scene.

As to methods of planting the little trees, the following suggestions may prove of value. As soon as the plants are received, they should be taken from the box and dipped in a thick puddle of water and loam. The roots must be thoroughly covered with the mud. Then the bundles into which the little trees are tied should be loosened and the trees placed in a trench dug on a slant. The dirt should be placed over the roots and the exposed parts of the plants covered with brush or burlap to keep away the rays of the sun.

When ready for planting, a few plants are dug up, set in a pail with thin mud at the bottom and carried to the place of planting. The most economical method of planting is for one man to make the holes with a mattock. These holes are made about a foot in diameter, by scraping off the sod with the mattock and then digging a little hole in the dirt underneath. A second man follows with a pail of plants and sets a single plant in this hole with his hands, see Fig. 129, making sure that the roots are straight and spread out on the bottom of the hole. The dirt should then be packed firmly around the plant and pressed down with the foot.

Improvement by cutting: The removal of certain trees in a grove is often necessary to improve the quality of the better trees, increase their growth, make the place accessible, and enhance its beauty. Cutting in a wooded area should be confined to suppressed trees, dead and dying trees and trees badly infested with insects and disease. In case of farm woodlands, mature trees of market value may be cut, but in parks and on private estates these have a greater value when left standing. The cutting should leave a clean stand of well-selected specimens which will thrive under the favorable influence of more light and growing space. Considerable care is required to prevent injury to the young trees when the older specimens are cut and hauled out of the woods. The marking of the trees to be removed can best be done in summer when the dead and live trees can be distinguished with ease and when the requisite growing space for each tree can be judged better from the density of the crowns. The cutting, however, can be done most advantageously in winter.

Immediately after cutting all diseased and infested wood should be destroyed. The sound wood may be utilized for various purposes. The bigger logs may be sold to the local lumber dealers and the smaller material may be used for firewood. The remaining brush should be withdrawn from the woodlot to prevent fire during the dry summer months.

In marking trees for removal, a number of considerations are to be borne in mind besides the elimination of dead, diseased and suppressed trees. When the marker is working among crowding trees of equal height, he should save those that are most likely to grow into fine specimen trees and cut out all those that interfere with them. The selection must also favor trees which are best adapted to the local soil and climatic conditions and those which will add to the beauty of the place. In this respect the method of marking will be different from that used in commercial forestry, where the aim is to net the greatest profit from the timber. In pure forestry practice, one sees no value in such species as dogwood, ironwood, juneberry, sumac and sassafras, and will therefore never allow those to grow up in abundance and crowd out other trees of a higher market value. But on private estates and in park woodlands where beauty is an important consideration, such species add wonderful color and attractiveness to the forest scene, especially along the roads and paths, and should be favored as much as the other hardier trees. One must not mark too severely in one spot or the soil will be dried out from exposure to sun and wind. When the gaps between the trees are too large, the trees will grow more slowly and the trunks will become covered with numerous shoots or suckers which deprive the crowns of their necessary food and cause them to "die back." Where the trees are tall and slim or on short and steep hillsides, it is also important to be conservative in marking in order that the stand may not be exposed to the dangers of windfall. No hard-and-fast rule can be laid down as to what would constitute a conservative percentage of trees to cut down. This depends entirely on the local conditions and on the exposure of the woodlot. But in general it is not well to remove more than twenty per cent of the stand nor to repeat the cutting on the same spot oftener than once in five or six years. The first cutting will, of course, be the heaviest and all subsequent cuttings will become lighter and lighter until the woodlot is put in good growing condition. On private estates and parks, where beauty is the chief aim, the woodland should be kept as natural, informal and as thick as possible. Where the woodland is cut up by many paths and drives, density of vegetation will add to the impression of depth and distance.

Protection: This subject has already been discussed considerably in the previous study on Forestry, and here it becomes necessary merely to add a few suggestions with special reference to private and park woodlands.

Guarding woodlands from *fire* is the most important form of protection. Surface fires are very common on small woodland holdings and the damage done to the standing vegetation is generally underestimated. An ordinary ground or surface fire on a woodland area will burn up the leaf-litter and vegetable mold, upon which the trees depend so much for food and moisture, and will destroy the young seedlings on the ground. Where the fire is a little more severe, the older trees are badly wounded and weakened and the younger trees are frequently killed outright. Insects and disease find these trees an easy prey, and all related forest conditions commence to deteriorate.

Constant watchfulness and readiness to meet any emergency are the keynote of effective fire protection. Notices similar to the one shown in Fig. 143 often help to prevent fires. It is also helpful to institute strict rules against dropping lighted matches or tobacco, or burning brush when the ground is very dry, or leaving smouldering wood without waiting to see that the fire is completely out. There should be many roads and foot-paths winding through the woodland in order that they may serve as checks or "fire lanes" in time of fire. These roads and paths should be kept free from brush and leaves and should be frequently patrolled. When made not too wide, unpretentious and in conformity with the natural surroundings, such drives and paths can become a very interesting feature of the place, winding through the woodland, exposing its charms and affording opportunity for pleasant driving and walking. The borders of the paths can be given special attention by placing the more beautiful native shrubs in prominent positions where they can lend increased attractiveness.

In case of fire, it should be possible to call for aid by telephone directly from the woodland and to find within easy reach the tools necessary to combat fire. It is also important to obtain the co-operation of one's neighbors in protecting the adjoining woodlands, because the dangers from insects, disease and fire threatening one bit of woodland area are more or less dependent upon the conditions in the adjoining woodland.

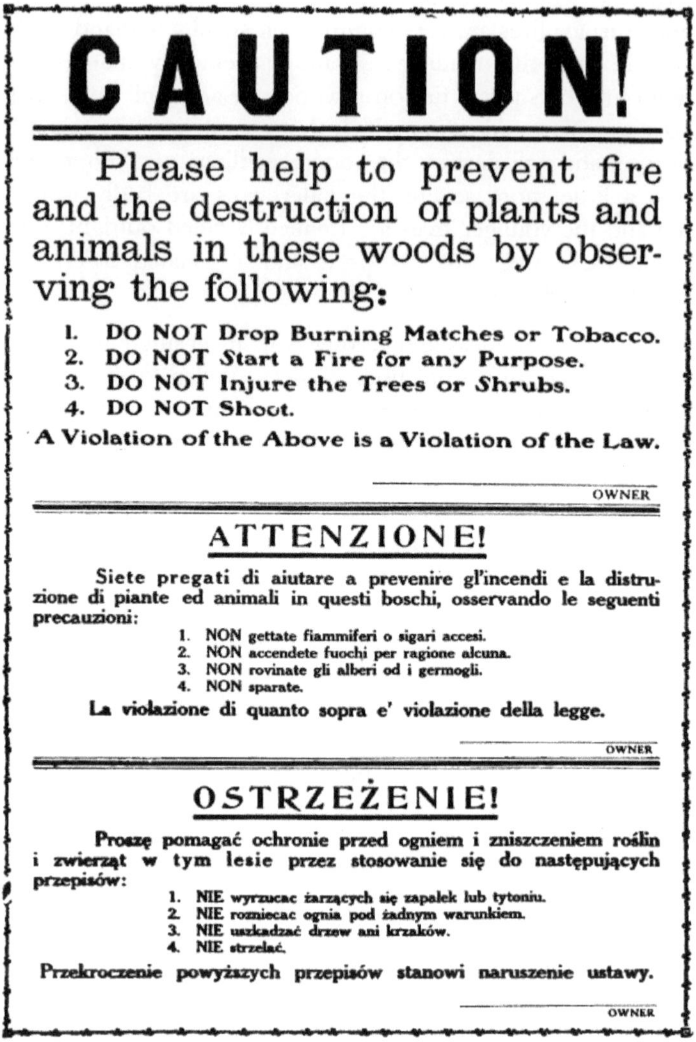

CAUTION!

Please help to prevent fire and the destruction of plants and animals in these woods by observing the following:

1. DO NOT Drop Burning Matches or Tobacco.
2. DO NOT Start a Fire for any Purpose.
3. DO NOT Injure the Trees or Shrubs.
4. DO NOT Shoot.

A Violation of the Above is a Violation of the Law.

OWNER

ATTENZIONE!

Siete pregati di aiutare a prevenire gl'incendi e la distruzione di piante ed animali in questi boschi, osservando le seguenti precauzioni:

1. NON gettate fiammiferi o sigari accesi.
2. NON accendete fuochi per ragione alcuna.
3. NON rovinate gli alberi od i germogli.
4. NON sparate.

La violazione di quanto sopra e' violazione della legge.

OWNER

OSTRZEŻENIE!

Proszę pomagać ochronie przed ogniem i zniszczeniem roślin i zwierząt w tym lesie przez stosowanie się do następujących przepisów:

1. NIE wyrzucac żarzących się zapałek lub tytoniu.
2. NIE rozniecac ognia pod żadnym warunkiem.
3. NIE uszkadzać drzew ani krzaków.
4. NIE strzelać.

Przekroczenie powyższych przepisów stanowi naruszenie ustawy.

OWNER

FIG. 143.—Poster Suitable for Private Woodlands and Forest Parks. The translations in Italian and Polish have been used by the writer in this particular instance to meet the local needs.

As to other forms of protection, passing mention may be made of the importance of keeping out cattle, sheep and hogs from the woods, of eliminating all insects and disease, of keeping the ground free from brush and other inflammable material, of retaining on the ground all fallen leaves and keeping the forest well stocked with little trees and shrubs.

Forest lands may be exempted from taxation: In New York and other States there exists a State law providing for exemption or reduction in taxes upon lands which are planted with forest trees or maintained as wooded areas. The object of the law is to encourage home forestry and to establish fairness in the agricultural land-tax law by placing forest lands in the same category with other crop-producing lands. For detailed information and a copy of the law, one should address the local State Forestry Commission.

Chapter VIII

Our Common Woods: Their Identification, Properties and Uses

Woods have different values for various practical purposes because of their peculiarities in structure. A knowledge of the structural parts of wood is therefore necessary as a means of recognizing the wood and of determining why one piece is stronger, heavier, tougher, or better adapted for a given service than another.

Structure of wood: If one examines a cross-section of the bole of a tree, he will note that it is composed of several distinct parts, as shown in Fig. 145. At the very center is a small core of soft tissue known as the *pith*. It is of much the same structure as the pith of cornstalk or elder, with which all are familiar. At the outside is the *bark*, which forms a protective covering over the entire woody system. In any but the younger stems, the bark is composed of an inner, live layer, and an outer or dead portion.

Between the pith at the center and the bark at the outside is the wood. It will be noted that the portion next to the bark is white or yellowish in color. This is the *sapwood*. It is principally through the sapwood that the water taken in by the roots is carried up to the leaves. In some cases the sapwood is very thin and in others it is very thick, depending partly on the kind of tree, and partly on its age and vigor. The more leaves on a tree the more sapwood it must have to supply them with moisture.

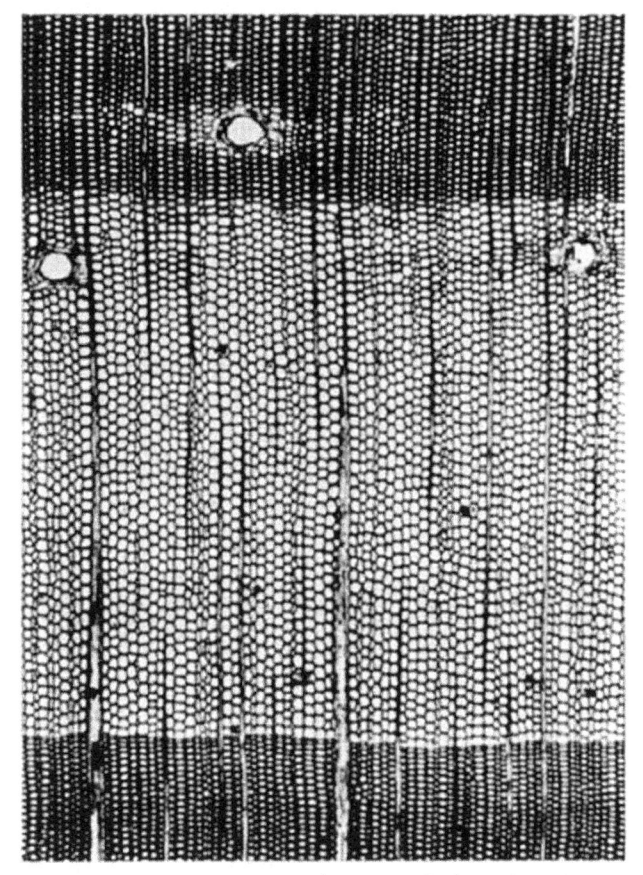

FIG. 144.—Pine Wood. (Magnified 30 times.)

Very young trees are all sapwood, but, as they get older, part of the wood is no longer needed to carry sap and it becomes *heartwood*. Heartwood is darker than the sapwood, sometimes only slightly, but in other instances it may vary from a light-brown color to jet black. It tends to fill with gums, resins, pigments and other substances, but otherwise its structure is the same as that of the sapwood.

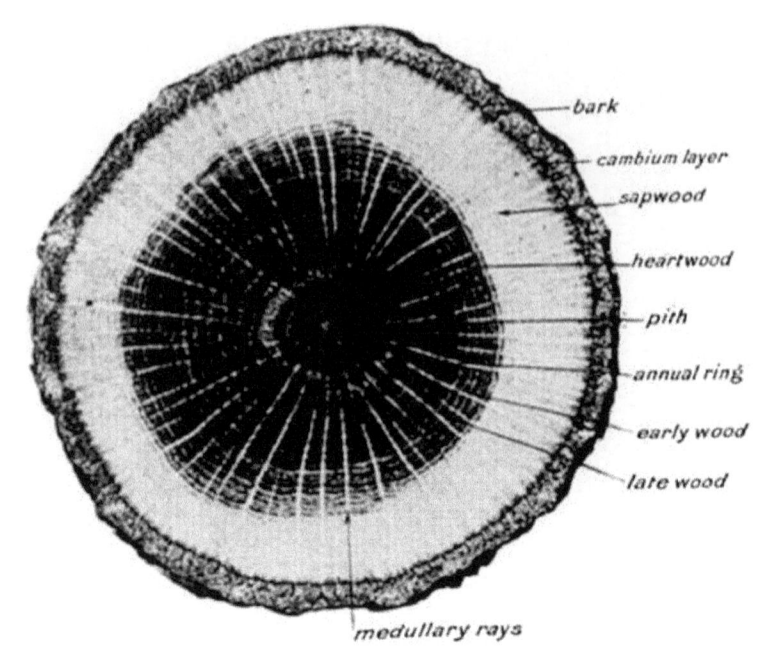

FIG. 145.—Cross-section of Oak.

The wood of all our common trees is produced by a thin layer of cells just beneath the bark, the *cambium*. The cambium adds new wood on the outside of that previously formed and new bark on the inside of the old bark. A tree grows most rapidly in the spring, and the wood formed at that time is much lighter, softer and more porous than that formed later in the season, which is usually quite hard and dense. These two portions, known as *early wood* or spring wood, and *late wood* or summer wood, together make up one year's growth and are for that reason called *annual rings*. Trees such as palms and yucca do not grow in this way, but their wood is not important enough in this country to warrant a description.

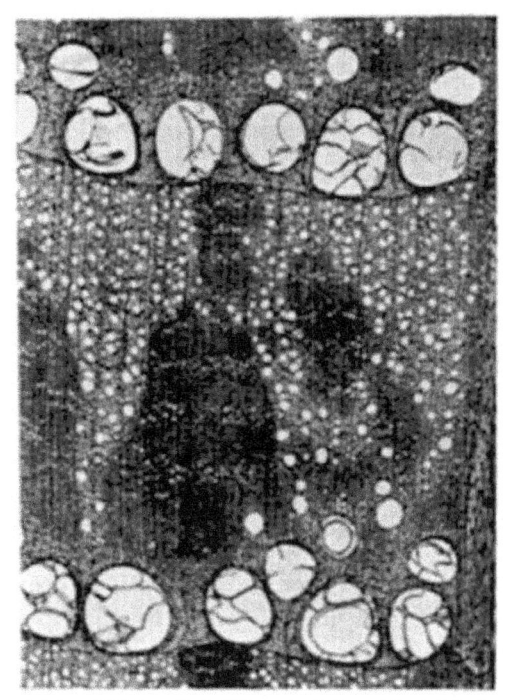

FIG. 146.—White Oak Wood. (Magnified 20 times.)

If the end of a piece of oak wood is examined, a number of lines will be seen radiating out toward the bark like the spokes in a wheel. These are the *medullary rays*. They are present in all woods, but only in a few species are they very prominent to the unaided eye. These rays produce the "flakes" or "mirrors" that make quartersawed (radially cut) wood so beautiful. They are thin plates or sheets of cells lying in between the other wood cells. They extend out into the inner bark.

While much may be seen with the unaided eye, better results can be secured by the use of a good magnifying glass. The end of the wood should be smoothed off with a very sharp knife; a dull one will tear and break the cells so that the structure becomes obscured. With any good hand lens a great many details will then appear which before were not visible. In the case of some woods like oak, ash, and chestnut, it will be found that the early wood contains many comparatively large openings, called *pores*, as shown in Figs. 146 and 147. Pores are cross-sections of vessels which are little tube-like elements running throughout the tree. The vessels are water carriers. A wood with its large pores collected into one row or in a single band is said to be *ring-porous*. Fig. 146 shows such an arrangement. A wood with its pores scattered throughout the year's growth instead of collected in a ring is *diffuse-porous*. Maple, as shown in Fig. 152, is of this character.

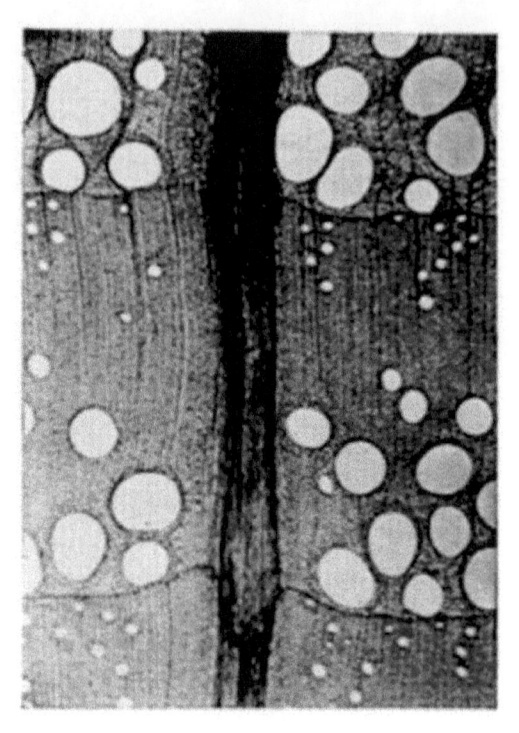

FIG. 147.—Example of the Black Oak Group. (Quercus coccinea.) (Magnified 20 times.)

All of our broadleaf woods are either ring-porous or diffuse-porous, though some of them, like the walnut, are nearly half way between the two groups.

If the wood of hickory, for example, be examined with the magnifying lens, it will be seen that there are numerous small pores in the late wood, while running parallel with the annual rings are little white lines such as are shown in Fig. 149. These are lines of *wood parenchyma*. Wood parenchyma is found in all woods, arranged sometimes in tangential lines, sometimes surrounding the pores and sometimes distributed over the cross-section. The dark, horn-like portions of hickory and oak are the *woodfibers*. They give the strength to wood.

In many of the diffuse-porous woods, the pores are too small to be seen with the unaided eye, and in some cases they are not very distinct even when viewed with a magnifier. It is necessary to study such examples closely in order not to confuse them with the woods of conifers.

The woods of conifers are quite different in structure from broadleaf woods, though the difference may not always stand out prominently. Coniferous woods have no pores, their rays are always narrow and inconspicuous, and wood parenchyma is never prominent. The woods of the pines, spruces, larches, and Douglas fir differ from those of the other conifers in having *resin ducts*, Fig. 144. In pines these are readily visible to the naked eye, appearing as resinous dots on cross-sections and as pin scratches or dark lines on longitudinal surfaces. The presence or absence of resin ducts is a very important feature in identifying woods, hence it is very important to make a careful search for them when they are not readily visible.

How to identify a specimen of wood: The first thing to do in identifying a piece of wood is to cut a smooth section at the end and note (without the magnifier) the color, the prominence of the rays and pores, and any other striking features. If the pores are readily visible, the wood is from a broadleaf tree; if the large pores are collected in a ring it belongs to the ring-porous division of the broadleaf woods. If the rays are quite conspicuous and the wood is hard and heavy, it is oak, as the key given later will show. Close attention to the details of the key will enable one to decide to what group of oaks it belongs.

In most cases the structure will not stand out so prominently as in oak, so that it is necessary to make a careful study with the hand lens. If pores appear, their arrangement, both in the early wood and in the late wood, should be carefully noted; also whether the pores are open or filled with a froth-like substance known as *tyloses*. Wood parenchyma lines should be looked for, and if present, the arrangement of the lines should be noted.

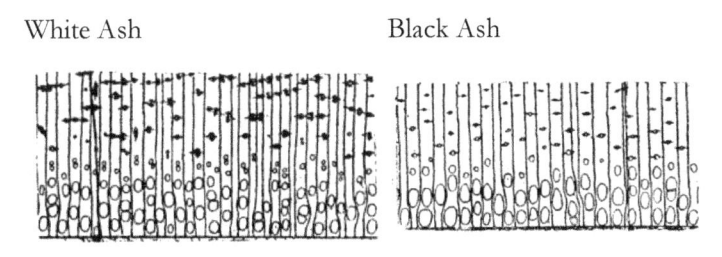

White Ash   Black Ash

FIG. 148.—(Magnified about 8 times.)

If no pores appear under the magnifying lens, look closely for resin ducts. If these are found, note whether they are large or small, numerous or scattered, open or closed, lighter or darker than the wood. Note also whether the late wood is very heavy and hard, showing a decided contrast to the early wood, or fairly soft and grading into the early wood without abrupt change. Weigh the piece in your hand, smell a fresh-cut surface to detect the odor, if any, and taste a chip to see if anything characteristic is discoverable. Then turn to the following key:

KEY

I. WOODS WITHOUT PORES—CONIFERS OR SO-CALLED "SOFTWOODS"

A. Woods with resin ducts.

1. **Pines.** Fig. 144. Resin ducts numerous, prominent, fairly evenly distributed. Wood often pitchy. Resinous odor distinct. Clear demarcation between heart and sapwood. There are two groups of pines—soft and hard.

(*a*) *Soft Pines.* Wood light, soft, not strong, even-textured, very easy to work. Change from early wood to late wood is gradual and the difference in density is not great.

(*b*) *Hard Pines.* Wood variable but typically rather heavy, hard and strong, uneven textured, fairly easy to work. Change from early wood to late wood is abrupt and the difference in density and color is very marked, consequently alternate layers of light and dark wood show. The wood of nearly all pines is very extensively employed in construction work and in general carpentry.

2. **Douglas fir.** Resin ducts less numerous and conspicuous than in the pines, irregularly distributed, often in small groups. Odorless or nearly so. Heartwood and sapwood distinct. The wood is of two kinds. In one the growth rings are narrow and the wood is rather light and soft, easy to work, reddish yellow in color; in the other the growth rings are wide, the wood is rather hard to work, as there is great contrast between the weak early wood and the very dense late wood of the annual rings.

Douglas fir is a tree of great economic importance on the Pacific Coast. The wood is much like hard pine both in its appearance and its uses.

3. **Spruces.** Resin ducts few, small, unevenly distributed; appearing mostly as white dots. Wood not resinous; odorless. The wood is white or very light colored with a silky luster and with little contrast between heart and sapwood. It is a great deal like soft pine, though lighter in color and with much fewer and smaller resin ducts. The wood is used for construction, carpentry, oars, sounding boards for musical instruments, and paper pulp.

4. **Tamarack.** Resin ducts the same as in the spruces. The color of the heartwood is yellowish or russet brown; that of the distinct sapwood much lighter. The wood is considerably like hard pine, but lacks the resinous odor and the resin ducts are much fewer and smaller.

The wood is used largely for cross-ties, fence posts, telegraph and telephone poles, and to a limited extent for lumber in general construction.

B. **Woods without resin ducts.**

1. **Hemlock.** The wood has a disagreeable, rancid odor, is splintery, not resinous, with decided contrast between early and late wood. Color light brown with a slight tinge of red, the heart little if any darker than the sapwood. Hemlock makes a rather poor lumber which is used for general construction, also for cross-ties, and pulp.

2. **Balsam fir.** Usually odorless, not splintery, not resinous, with little contrast between early and late wood. Color white or very light brown with a pinkish hue to the late wood. Heartwood little if any darker than the sapwood. Closely resembles spruce, from which it can be distinguished by its absence of resin ducts.

The wood is used for paper pulp in mixture with spruce. Also for general construction to some extent.

3. **Cypress.** Odorless except in dark-colored specimens which are somewhat rancid. Smooth surface of sound wood looks and feels greasy or waxy. Moderate contrast between early and late wood. Color varies from straw color to dark brown, often with reddish and greenish tinge. Heartwood more deeply colored than the sapwood but without distinct boundary line.

Wood used in general construction, especially in places where durability is required; also for shingles, cooperage, posts, and poles.

4. **Red Cedar.** Has a distinct aromatic odor. Wood uniform-textured; late wood usually very thin, inconspicuous. Color deep reddish brown or purple, becoming dull upon exposure; numerous minute red dots often visible under lens. Sapwood white. Red cedar can be distinguished from all the other conifers mentioned by the deep color of the wood and the very distinct aromatic odor.

Wood largely used for pencils; also for chests and cabinets, posts, and poles. It is very durable in contact with the ground.

Western red cedar is lighter, softer, less deeply colored and less fragrant than the common Eastern cedar. It grows along the Pacific Coast and is extensively used for shingles throughout the country.

5. **Redwood.** Wood odorless and tasteless, uniform-textured, light and weak, rather coarse and harsh. Color light cherry. Close inspection under lens of a small split surface will reveal many little resin masses that appear as rows of black or amber beads which are characteristic of this wood.

Redwood is confined to portions of the Pacific Coast. It is used for house construction, interior finish, tanks and flumes, shingles, posts, and boxes. It is very durable.

WOODS WITH PORES—BROADLEAF, OR SO-CALLED "HARDWOODS"

A. Ring-porous.

1. *Woods with a portion of the rays very large and conspicuous.*

Oak. The wood of all of the oaks is heavy, hard, and strong. They may be separated into two groups. The white oaks and the red or black oaks.

(*a*) *White oaks.* Pores in early wood plugged with tyloses, collected in a few rows. Fig. 146. The transition from the large pores to the small ones in the late wood is abrupt. The latter are very small, numerous, and appear as irregular grayish bands widening toward the outer edge of the annual ring. Impossible usually to see into the small pores with magnifier.

(*b*) *Red or black oaks.* Pores are usually open though tyloses may occur, Fig. 147; the early wood pores are in several rows and the transition to the small ones in late wood is gradual. The latter are fewer, larger and more distinct than in white oak and it is possible to see into them with a hand lens.

The wood of the oaks is used for all kinds of furniture, interior finish, cooperage, vehicles, cross-ties, posts, fuel, and construction timber.

2. *Woods with none of the rays large and conspicuous.*

(*a*) Pores in late wood small and in radial lines, wood parenchyma in inconspicuous tangential lines.

Chestnut. Pores in early wood in a broad band, oval in shape, mostly free from tyloses. Pores in late wood in flame-like radial white patches that are plainly visible without lens. Color medium brown. Nearly odorless and tasteless. Chestnut is readily separated from oak by its weight and absence of large rays; from black ash by the arrangement of the pores in the late wood; from sassafras by the arrangement of the pores in the late wood, the less conspicuous rays, and the lack of distinct color.

The wood is used for cross-ties, telegraph and telephone poles, posts, furniture, cooperage, and tannin extract. Durable in contact with the ground.

(*b*) Pores in late wood small, not radially arranged, being distributed singly or in groups. Wood parenchyma around pores or extending wing-like from pores in late wood, often forming irregular tangential lines.

1. **Ash.** Pores in early wood in a rather broad band (occasionally narrow), oval in shape, see Fig. 148, tyloses present. Color brown to white, sometimes with reddish tinge to late wood. Odorless and tasteless. There are several species of ash that are classed as white ash and one that is called black or brown ash.

(*a*) *White ash.* Wood heavy, hard, strong, mostly light colored except in old heartwood, which is reddish. Pores in late wood, especially in the outer part of the annual ring, are joined by lines of wood parenchyma.

(*b*) *Black ash.* Wood more porous, lighter, softer, weaker, and darker colored than white ash. Pores in late wood fewer and larger and rarely joined by tangential lines of wood parenchyma.

The wood of the ashes is used for wagon and carriage stock, agricultural implements, oars, furniture, interior finish, and cooperage. It is the best wood for bent work.

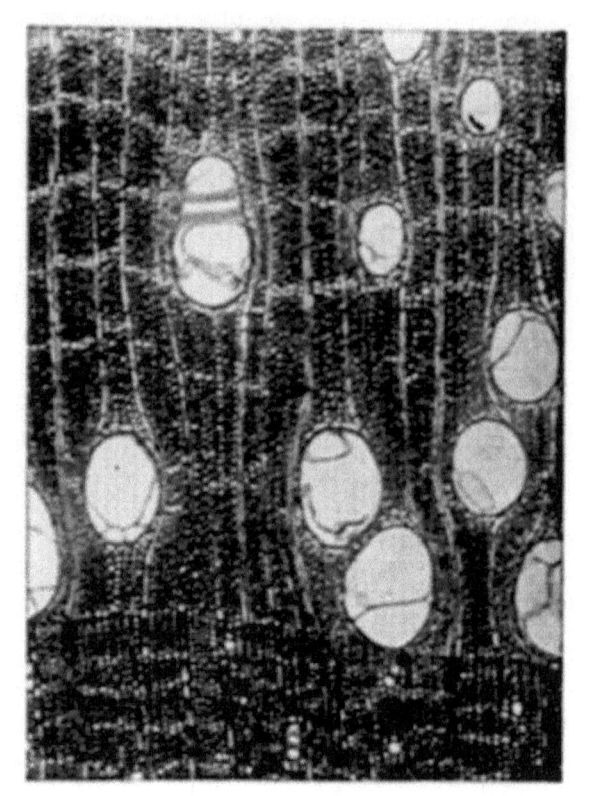

FIG. 149.—Hickory Wood. (Magnified 45 times.)

2. **Locust.** Pores in early wood in a rather narrow band, round, variable in size, densely filled with tyloses. Color varying from golden yellow to brown, often with greenish hue. Very thin sapwood, white. Odorless and almost tasteless. Wood extremely heavy and hard, cutting like horn. Locust bears little resemblance to ash, being harder, heavier, of a different color, with more distinct rays, and with the pores in late wood in larger groups.

The wood is used for posts, cross-ties, wagon hubs, and insulator pins. It is very durable in contact with the ground.

(*c*) Pores in late wood comparatively large, not in groups or lines. Wood parenchyma in numerous fine but distinct tangential lines.

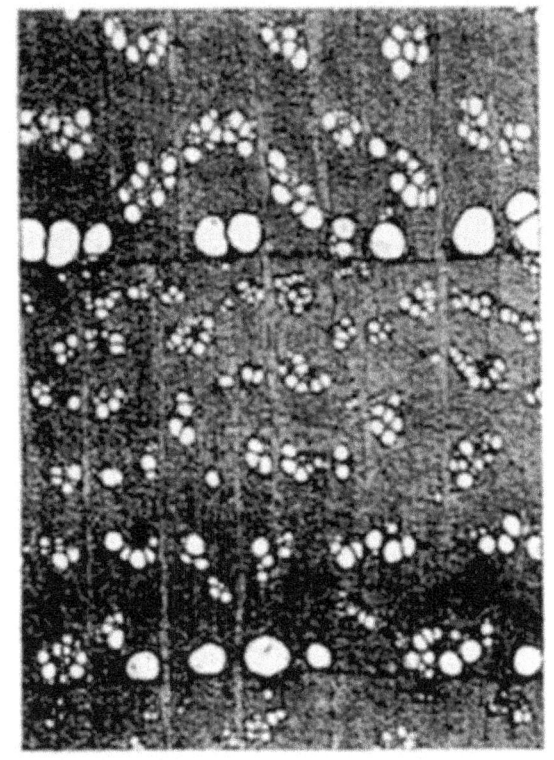

FIG. 150.—Elm. (Magnified 25 times.)

Hickory, Fig. 149. Pores in early wood moderately large, not abundant, nearly round, filled with tyloses. Color brown to reddish brown; thick sapwood, white. Odorless and tasteless. Wood very heavy, hard, and strong. Hickory is readily separated from ash by the fine tangential lines of wood parenchyma and from oak by the absence of large rays.

The wood is largely used for vehicles, tool handles, agricultural implements, athletic goods, and fuel.

(*d*) Pores in late wood small and in conspicuous wavy tangential bands. Wood parenchyma not in tangential lines.

Elm. Pores in early wood not large and mostly in a single row, Fig. 150 (several rows in slippery elm), round, tyloses present. Color brown, often with reddish tinge. Odorless and tasteless. Wood rather heavy and hard, tough, often difficult to split. The peculiar arrangement of the pores in the late wood readily distinguishes elm from all other woods except *hackberry*, from which it may be told by the fact that in elm the medullary rays are indistinct, while they are quite distinct in hackberry; moreover, the color of hackberry is yellow or grayish yellow instead of brown or reddish brown as in elm.

The wood is used principally for slack cooperage; also for hubs, baskets, agricultural implements, and fuel.

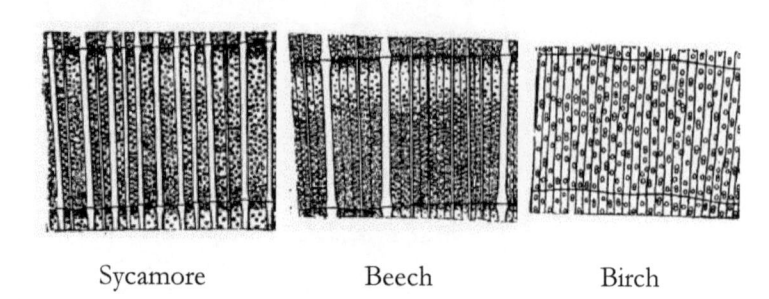

Sycamore   Beech   Birch

FIG. 151.—(Magnified about 8 times.)

B. Diffuse-porous.

1. *Pores varying in size from rather large to minute, the largest being in the early wood. Intermediate between ring-porous and diffuse-porous.*

Black Walnut. Color rich dark or chocolate brown. Odor mild but characteristic. Tasteless or nearly so. Wood parenchyma in numerous, fine tangential lines. Wood heavy and hard, moderately stiff and strong. The wood is used principally for furniture, cabinets, interior finish, moulding, and gun stocks.

2. *Pores all minute or indistinct, evenly distributed throughout annual ring.*

(*a*) With conspicuously broad rays.

1.   **Sycamore.** Fig. 151. Rays practically all broad. Color light brown, often with dark stripes or "feather grain." Wood of medium weight and strength, usually cross-grained, difficult to split.

The wood is used for general construction, woodenware, novelties, interior finish, and boxes.

2.   **Beech.** With only a part of the rays broad, the others very fine, Fig. 151. Color pale reddish brown to white; uniform. Wood heavy, hard, strong, usually straight-grained.

The wood is used for cheap furniture, turnery, cooperage, woodenware, novelties, cross-ties, and fuel. Much of it is distilled.

(*b*) Without conspicuously broad rays.

3. **Cherry.** Rays rather fine but very distinct. Color of wood reddish brown. Wood rather heavy, hard, and strong.

The wood is used for furniture, cabinet work, moulding, interior finish, and miscellaneous articles.

4. **Maple,** Fig. 152. With part of the rays rather broad and conspicuous, the others very fine. Color light brown tinged with red. The wood of the hard maple is very heavy, hard and strong; that of the soft maples is rather light, fairly strong. Maple most closely resembles birch, but can be distinguished from it through the fact that in maple the rays are considerably more conspicuous than in birch.

The wood is used for slack cooperage, flooring, interior finish, furniture, musical instruments, handles, and destructive distillation.

5. **Tulip-tree, yellow poplar or whitewood.** Rays all fine but distinct. Color yellow or brownish yellow; sapwood white. Wood light and soft, straight-grained, easy to work.

The wood is used for boxes, woodenware, tops and bodies of vehicles, interior finish, furniture, and pulp.

6. **Red or sweet gum.** Rays all fine but somewhat less distinct than in tulip tree. Color reddish brown, often with irregular dark streaks producing a "watered" effect on smooth boards; thick sapwood, grayish white. Wood rather heavy, moderately hard, cross-grained, difficult to work.

The best grades of figured red gum resemble Circassian walnut, but the latter has much larger pores unevenly distributed and is less cross-grained than red gum.

The wood is used for finishing, flooring, furniture, veneers, slack cooperage, boxes, and gun stocks.

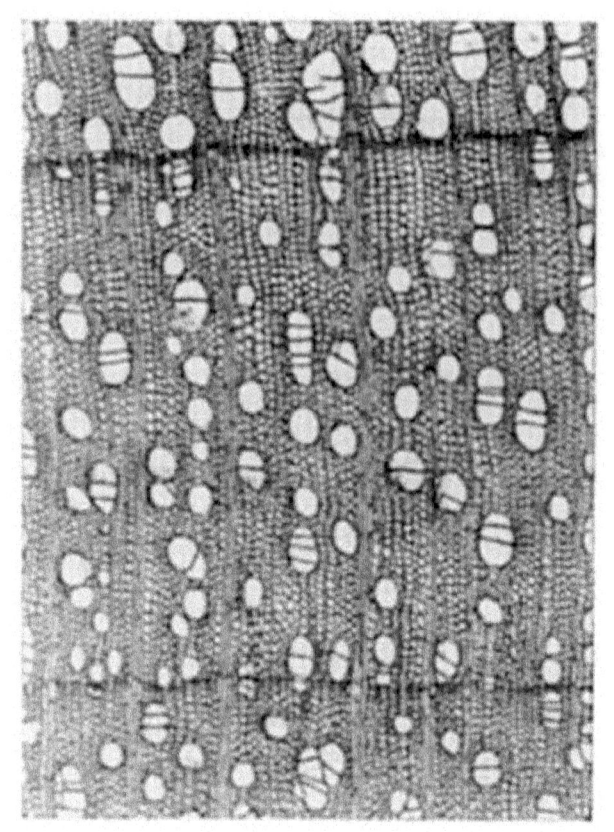

FIG. 152.—Maple. (Magnified 25 times.)

7. **Black or sweet birch,** Fig. 151. Rays variable in size but all rather indistinct. Color brown, tinged with red, often deep and handsome. Wood heavy, hard, and strong, straight-grained, readily worked. Is darker in color and has less prominent rays than maple.

The wood is used for furniture, cabinet work, finishing, and distillation.

8. **Cottonwood.** Rays extremely fine and scarcely visible even under lens. Color pale dull brown or grayish brown. Wood light, soft, not strong, straight-grained, fairly easy to work. Cottonwood can be separated from other light and soft woods by the fineness of its rays, which is equaled only by willow, which it rather closely resembles. The wood is largely used for boxes, general construction, lumber, and pulp.

How to judge the quality of wood: To know the name of a piece of wood means, in a general way, to know certain qualities that are common to all other pieces of wood of that species, but it does not explain the special peculiarities of the piece in question or why that particular piece is more suitable or unsuitable for a particular purpose than another piece of the same species. The mere identification of the wood does not explain why a particular piece is tougher, stronger or of darker color than another piece of the same species or even of the same tree. The reason for these special differences lies in the fact that wood is not a homogeneous material like metal. Within the same tree different parts vary in quality. The heartwood is generally heavier and of deeper color than the sapwood. The butt is superior to the top wood, and the manner in which the wood was sawed and dried will affect its quality. Knots, splits, checks, and discoloration due to incipient decay are defects worth considering. Wood that looks lusterless is usually defective, because the lack of luster is generally due to disease. Woods that are hard wear best. Hardness can be determined readily by striking the wood with a hammer and noting the sound produced. A clear, ringing sound is a sign of hardness. The strength of a piece of wood can be judged by its weight after it is well dried. Heavy woods are usually strong. A large amount of late wood is an indication of strength and the production of a clear sound when struck with a hammer is also an evidence of strength.

Chapter IX

An Outdoor Lesson on Trees

The importance of nature study in the training of the child is now well recognized. The influences of such study from the hygienic, moral and æsthetic point of view are far reaching and cannot be expressed in dollars and cents. In his association with nature, the child is led to observe more closely and to know and to be fond of what is truly beautiful in life—beautiful surroundings, beautiful thoughts and beautiful deeds. He is inspired with reverence for law, order and truth because he sees it constantly reflected in all works of nature. The social instinct is highly developed and even the parents are often bettered through the agency of their children.

The only way, however, to study nature—especially plants—is to study it out of doors. Our present tendency to gather in cities demands the upbuilding influences of trips into the open in order to equip the child mentally and physically to face the world and its work with the strength and tenacity characteristic of the country-bred. Moreover, the study of objects rather than books is an axiom in modern education and here, too, we can readily see that the best way to study trees is to take the pupil to the trees. Such studies are more lasting than book study because they emphasize the spirit and the goal rather than the petty facts.

Educators and parents are now recognizing the value of outdoor trips for their children and are beginning to indulge in them quite frequently. In many instances teachers about to take out their children for a day have inquired of the writer how to go about giving a general field lesson when they reached the park or woodland. The purpose of this chapter is to answer such a question and yet it is evident that it cannot be answered completely. What to observe out doors and how to present one's impressions is a broad question and varies with the knowledge and ability of the teacher as well as with the age and experience of the children. The how and the what in nature study is of greater import than the hard, dry facts and that must be left entirely to the teacher. A few suggestions, however, may not be amiss:

1. General observations with a view to character building: First of all it is important to remember that the great value of all tree and nature study is the inculcation in the minds of the children of an appreciation and love for the beautiful. Inspiring them to *love* trees generally means more than teaching them to *know* trees. Mere facts about trees taught in an academic way are often no more lasting than the formulæ in trigonometry which most of us have long ago forgotten. The important thing is that permanent results be left and nothing else will produce such lasting impressions as the study of trees out of doors.

FIG. 153.—Trees Have Individuality.

General observations about trees can be made by pointing out the beauty and character of the individual forms and branching, their harmony in their relations to each other as factors of a beautiful composition and the wealth of shades and colors in their leaves, bark and flowers. Compare, for instance, the intricate ramification of an American elm with the simple branching of a sugar maple, the sturdiness of a white oak with the tenderness of a soft maple, the wide spread of a beech with the slender form of a Lombardy poplar, the upward pointing branches of a gingko with the drooping form of a weeping willow. At close range, each of these trees reveals itself as an individual with a character quite its own. At little distance you may see them grouped together, subordinating their individuality and helping to blend into a beautiful composition with a character all its own.

There is nothing more inspiring than the variety of greens in the spring foliage, the diversity of color in the spring blossoms and the wonderful display of autumnal tints offered by the sweet gum, sassafras, dogwood, black gum, red maple, sugar maple, scarlet oak, blue beech, sorrel tree, ash and gingko. The white bark of the gray birch, the dark bark of the black oak, the gray of the beech, the golden yellow of the mulberry and the mottled bark of the sycamore are interesting comparisons. The smooth bark of the mockernut hickory contrasts greatly with the shaggy bark of the shagbark hickory—members of the same family and yet how different. A wonderful opportunity is thus offered for a comparative study of human nature—individuality and community life, all reflected in trees.

With this preliminary study and with the addition of some remarks on the value of trees as health givers and moral uplifters, the child is interested and attracted. The lesson so far has attained its aim.

2. Specific observations with a view to training the observative powers: The child's training in closeness of observation and scientific precision may be the next consideration. His enthusiasm will now prompt him to lend his interest for greater detail. We can teach him to recognize a few of the common trees by their general characters—an American elm by its fan-shaped form, a gray birch by its white bark, a white pine by the five needles to each cluster, a horsechestnut by its opposite branching and big sticky bud and a willow by its drooping habit. After that we may introduce, if the age of the pupils justifies, more details extending to greater differences which distinguish one species from another.

The lesson might continue by pointing out the requirements of trees for water and light. Find a tree on some slope where the roots are exposed and another which is being encroached upon by its neighbor, and show how in one case the roots travel in search of water and food and in the other the branches bend toward the light, growing more vigorously on that side. Compare the trees on the open lawn with those in the grove and show how those in the open have grown with branches near the ground while those in the woodland are slender, tall and free from branches to some distance above the ground. Point out the lenticels on the bark of birch and sweet cherry trees and explain how trees breathe. Compare this process with that of the human body. You may now come across an old stump and here you can point out the structure of the wood—the sapwood, cambium and bark. You can illustrate the annual rings and count the age of the tree. At another point you may find a tree with a wound or bruised bark and here you can readily make a closer study of the cambium layer and its manner of growth.

The adaptation of plants to the seasonal changes opens another interesting field of study for beginners. If the season is the fall or winter, note how the trees have prepared themselves for the winter's cold by terminating the flow of sap, by dropping their leaves too tender to resist the winter's cold, and by covering their buds with scales lined with down on the inside. Observe how the insects have spun for themselves silken nests or remain preserved in the egg state over the winter. If the season is spring or summer the opposite may be noted. See how everything turns to life; how the buds are opening, the leaves emerging, the sap running, seeds germinating and flowers blooming.

The soil conditions on the lawn and in the grove furnish another interesting feature of comparison and study. In the grove, you can demonstrate the decomposition of the fallen leaves, the formation of humus and its value to the tree. The importance of the forest soil as a conservator of water and its relation to stream flow and soil erosion can be brought out at this juncture. An eroded bank and a slope covered with trees and shrubs would provide excellent models for this study. A consideration of the economic value of the trees would also be in place.

3. Civic lessons reflected in trees: The community life of trees in the grove, their growth, struggles for light and food and their mutual aid can be brought out and compared with the community life among people. The trees may here be seen struggling with each other for light and food, forcing each other's growth upward, some winning out and developing into stalwart and thrifty specimens and others becoming suppressed or entirely killed. On the other hand they may be seen helping each other in their community growth by protecting each other from windfall and by contributing to the fertility of the forest soil in dropping their leaves and shading the ground so that these fallen leaves may decompose readily.

FIG. 154.—Trees also Grow in Communities.

4. Enemies of trees: An old stump or tree may be seen crumbling away under the influence of fungi and here the children may be shown the effects of tree diseases both as destroyers of life and as up-builders, because fungi turn to dust the living trees and build up others by furnishing them with the decomposed wood matter.

Insects too, may be invading the old dead tree, and something of their nature, habits and influences may be gone into. They may be shown as wood borers, leaf eaters, or sap suckers, all injurious to the tree. On the other hand they may be shown as seed disseminators and as parasites on other injurious insects; all benefactors.

Forest fires as an enemy of trees might be touched upon by noting how easily the leaves may be ignited and a surface fire started when the season is dry. Top and ground fires emanating from surface fires can then be readily explained.

FIG. 155.—Trees Blend Together to Form a Beautiful Composition.

5. Expression: The pupils have by this time been taught to feel the beautiful, to observe carefully and to reason intelligently and they may now be trained to express themselves properly. This may be accomplished by asking them to remember their observations and to write about them in the classroom. The lesson may be supplemented with effective reading about trees and forests. Interesting reading matter of this sort can be found in abundance in children's readers, in special books on the subject and in Arbor Day Manuals published by the various State Education Departments.

6. Preparation: In order to save time looking for objects of interest and for the purpose of correlating the various observations so that all will follow in orderly sequence, it is well for the teacher or leader to go over the ground beforehand and note the special features of interest. The various topics can then be given some thought and a brief synopsis can be drawn up to serve as a memorandum and guide on the trip.

It is also well to be provided with a hatchet to cut into some decayed stump, a trowel to dig up the forest soil, a knife for cutting off twigs and a hand reading glass for examining the structural parts of the various objects under observation. A camera is always a valuable asset because the photographs hung in the classroom become records of great interest to all participants.

7. Suggestions for forming tree clubs: A good way to interest children in trees and nature study is to form, among them, a Tree Club. The idea has been fully developed in Brooklyn, N. Y., Newark, N. J., and other cities and consists of forming clubs of children in the public schools and private institutions for the purpose of interesting them in the trees around their school and their homes. The members of these clubs are each given the tree warden's badge of authority and assigned to some special duty in the preservation of the local trees. A plan of study and of outdoor trips is laid out for them by their director and at stated periods they are given illustrated lectures on trees and taken to the neighboring parks or woodlands.